3rd

PCT
PEST CONTROL TECHNOLOGY

TECHNICIAN'S HANDBOOK

A Guide To Pest Identification & Management

By Dr. Richard Kramer

Editor
Jeff Fenner

Contributing Editor
Dan Moreland

Production Manager
Helen Duerr

Art Director
Carolyn Antl

Cover Photograph
Roger Mastroianni

Copyright 1999 by G.I.E. Inc., Publishers. Reprinted in 2002.

rmation requests should be directed to: PCT Technician's Handbook, 4012 Bridge Ave., Cleveland, OH 113, 216/961-4130 or www.pctonline.com.

ISBN: 1-883751-08-X
Library of Congress Catalog Card No.: 97-077870

Dedication

TO

The many service technicians who are the backbone of the structural pest management industry. It is through their expertise and commitment that the pest management industry can be proud of the services and benefits it provides to society.

AND

Paul M. Kramer, Sr., my father, who taught me the values of integrity and honesty in business, and guided me down the right path in life and this industry.

AND

The memory of my mother, Marjorie Kramer, who always stessed the value of education and who wanted me to become a medical doctor.

AND

Bonnie, my wife, who spent many long hours editing this book and translating my thoughts into the English language. — *Dr. Richard Kramer*

Acknowledgements

The PCT Media Group would like to acknowledge and thank the following companies and individuals for their assistance in providing illustrations and photographs for the PCT Technician's Handbook: Vopak; Aventis Environmental Science; Dow AgroSciences, Syngenta Professional Products; ABC Pest Control Dallas/Austin, Texas; and Western Exterminator Co., Irvine, Calif. — *Jeff Fenner*

TABLE OF CONTENTS

Pest Profiles

Ants

Blood Feeders

• = Color Photograph Available

• = Color Photograph Available

• = Color Photograph Available

CHAPTER 1
Pest Management and Society

Introduction

ROLE OF THE TECHNICIAN

Pest management technicians have three primary responsibilities — as ambassadors for the industry, public educators, and service providers. The value of the structural pest management industry to the immediate customer as well as to consumers in general is often not communicated. Today's technicians must be armed with facts and prepared to communicate to both individuals and groups the vital services that are provided by the pest manage-

The professional technician projects a positive image.

ment industry, specifically, the protection of health, property, food, and the environment.

Pest management technicians often find themselves on the defensive, facing continuous pressure to significantly reduce or eliminate pesticide use. This was brought to the forefront as environmentalist groups and individuals expounded a different view of pesticides and their value to society. They have had justifiable concerns and have succeeded where the pest management industry has failed because they are able to capture the attention of consumers via media and their educational efforts.

The primary task of the structural pest management industry once was to spray pesticides. During the past decade, however, the industry gradually evolved, shifting to both integrated pest management (IPM) and reduced risk pest management strategies. These processes begin with the identification of pest problems and are followed by the implementation of a comprehensive pest management plan that may involve application of pesticides. Today baseboard spraying is "out," and techniques that minimize the use of pesticides, e.g., crack and crevice applications as well as baits, insect growth regulators, and other approaches to pest management have become the norm.

Most consumers are unaware of these changes and so continue to expect an "exterminator" to walk in the door with a compressed air sprayer in hand. When pest management technicians take the time to explain new techniques and products to consumers, it is an investment in the future. Technicians should be prepared to answer tough questions regarding pest biology and behavior, justify the use of reduced risk techniques, and explain the benefits of the services they provide.

Understanding and anticipating consumer expectations are important responsibilities for everyone involved in the industry, but it is especially important for the pest management technician to explain how the process and results may differ depending on the type of pest management strategy to be employed, e.g., the use of residual pesticides and aerosols often results in immediate reduction of pest populations, whereas the use of baits and insect growth regulators requires more time. Failure to explain such information might lead to misunderstanding and dissatisfaction.

Pest management technicians should be familiar with the information on the product label and knowledgeable about all aspects of pesticide use when the pest management plan involves pesticide use. Some of the toughest questions are those regarding pesticide safety, risk of exposure during application, reentry time, health effects, etc. Vital to consumer education is the ability to communicate this knowledge when responding to such questions. Technicians who are knowledgeable and comfortable in their responses provide a positive image for their companies and for the pest management industry.

Technicians who are comfortable in their dealings with customers provide a positive image for the industry.

Technicians are obligated to be professional in their physical appearance as well as in their use of body language, i.e., they should be neatly attired in clean uniforms, and they, themselves, should be clean. They should be good listeners, responding honestly to customers' questions. If they don't know the answer to a question, they should find out and communicate the answer to the customer as soon as possible.

The customer's first impression of the pest management technician is crucial because it is also a first impression of the pest management company, and it may even be a first impression of the industry as a whole. Hopefully, the first impression will not be the last.

Pests are defined as plants and animals that cause trouble, annoyance, or discomfort to humans and/or other animals. Customers often define pests according to their threshold of acceptance. Most individuals find just one cockroach to be an unacceptable number, but a few people tolerate cockroaches as long as they don't see them.

Customers are intolerant of most insect and other animal pests because they are unsightly or are indicative of unsanitary conditions. However, the deleterious effects of pests on human health is the most important reason for utilization of pest management services. Pests affect human and animal health in various ways: as modifiers of commodities, pathogens, vectors of disease, and targets of control agents. In other words, pests affect human and animal health by consuming and contaminating food supplies; causing psychological stress; injecting venoms; invading body tissues; transmitting viral, bacterial, fungal, and other disease-causing organisms; and exposing them to the potential hazards of pesticides.

The most important responsibility of the pest management technician is to provide safe pest management services to the consuming public. The benefits provided to the public and the environment by modern pest management services are described below.

IMPACT ON FOOD

In addition to consumption of human and animal food, pests affect the physical characteristics of food, its palatability, seed viability, and nutritional value. Several investigators have estimated post-harvest food loss as a result of pests at 8% to 25% in developed countries such as the United States and as high as 70% in developing tropical countries.

This difference is affected by several contributing factors, i.e., climate, higher reproductive potential of the pest population, and inadequate storage and handling. However, the most significant factor is the availability and implementation of innovative pest management materials and practices which are available in the United States but are only just beginning to be used by technicians in developing countries.

The pest management technician plays a major role in protecting the food industry.

In addition, the expectations that consumers have for an abundant and wholesome food supply are a major influence on the demand for effective pest management services in the United States.

In addition to consumption of food products, pests affect food quality such as taste. Secretory and excretory products contaminate foods. For example, confused and red flour beetles impart a very distasteful quality to the products they infest. Other health hazards associated with pests in food products include human consumption of living (and fragments of) arthropods, e.g., the hastasetae. (i.e., barbed hairs) associated with dermestid larvae are known to occasionally cause severe gastric distress in young children. Consumption of materials of arthropod origin are known to cause severe allergic problems in sensitive individuals. Rats, mice, and birds also consume food products and contaminate them with their wastes.

The pest management industry plays a critical role in protecting food products, i.e., during transportation, storage, processing and manufacturing, distribution, retail sales, customer storage, and, ultimately, consumption.

IMPACT ON HEALTH

A variety of arthropod pests are pathogenic, i.e., they cause injury to humans and domestic animals by producing psychological stress, etiological (i.e., causal) agents, and invading living tissue. This is in contrast to disease-transmitting pests which carry disease-causing organisms. The impact of pathogenic pests on humans and domestic animals runs the gamut from pain and suffering associated with imaginary pests to death from anaphylactic reactions to bee, wasp, and ant venoms.

A variety of pests are considered to be direct pathogens because of their ability to induce allergic reactions in sensitive individuals. The induction of allergies is caused by antigenic materials, i.e., naturally-occurring materials which contain protein and are associated with the pests, such as their cuticles (i.e., exterior skeletons), excretions, pheromones (i.e., communication chemicals), etc. For example, the role of cockroaches in human allergies and asthma has been well documented.

While most arthropods that are found on humans, e.g., lice, fleas, chiggers, ticks, and other blood feeders, are considered to be surface grazers, several pests do invade human tissue. The most common invaders of human

tissue are mites, e.g., the hair follicle and scabies mites. Occasionally, fly maggots, i.e., larvae, are found to be infesting wounds on animals or humans.

Humans are no more than a free lunch serving as an important link in the survival of many blood-feeding arthropods, e.g., ticks, mosquitoes, lice, and biting flies. Although most of these blood-feeders consume very little, the sensitivity reactions to their bites, which can include severe itching and welts, is usually more uncomfortable.

Some of the most unfortunate victims of these pests are individuals suffering from delusions that pests infest their bodies (i.e., delusory parasitosis) or houses and who simply have a fear of insects (i.e., entomophobia). These conditions are aggravated by the media which exaggerates, often graphically, the effects certain pests have on humans.

The risk of customer exposure, especially children, to pesticides has come under close scrutiny from the regulatory community.

Oftentimes, these "pest problems" are traced to environmental conditions, hormonal changes, medications, psychological conditions and myriad other problems that produce the perception of a pest problem and/or the sensation of having been bitten. With this in mind, the pest management technician must determine if there actually is a pest problem.

In the summertime, as pest populations explode, the potential for arthropod-borne disease transmission (e.g. Lyme disease, encephalitis, malaria, and dengue) and vertebrate, i.e., zoonotic diseases, (e.g. rabies, and hantavirus), increases significantly. The incidence of pest-borne diseases as well as their impact on human health are increasing. Emerging pest-transmitted diseases which result in devastating health effects include the hantavirus and Lyme disease, and malaria which at one time had been eradicated from the United States.

Several underlying causes exist for this increasing threat to human health. Diseases that are associated with rodent populations and other reservoir animals are increasing due to human encroachment into the animals' native habitats. Urban sprawl has reached out into agricultural areas and native woodlands, placing people and structures within the animals' habitat and displacing the animals from their native environment. Faced with a reduction in harborage and food, the animals seek new resources within and around dwellings and other structures.

Lyme disease is increasing in many rural residential settings. If left untreated, this extremely debilitating disease can cause premature arthritic conditions, heart problems, and other systemic disorders. Lyme disease is transmitted by the blacklegged tick in the eastern United States and by the western blacklegged tick along the Pacific coast. Deer and deer mice are also involved in the disease cycle; the latter commonly invade homes in late fall and early winter. Human exposure occurs in backyards frequented by the mice and by hiking through tick-infested areas.

The deer mouse is the major rodent species associated with hantavirus pulmonary syndrome. In this case, the disease is not transmitted by an insect or arthropod but is the result of inhaling mouse droppings or urine-contaminated particles which contain the virus. Most exposures have occurred in areas where mouse populations have increased with a resultant accumulation of droppings and nesting materials. The mortality rate for individuals who contract this disease is approximately 66%. Although this disease is an occupational risk for service technicians, there have been no fatalities.

In contrast to suburban settings, the inner cities of large metropolitan areas are plagued with burgeoning rodent populations. Rodent population growth appears to be related to a deterioration of living conditions in these areas, breakdown in government services (i.e., especially sanitation services), and revitalization projects which often displace established rodent populations. Inner city inhabitants commonly observe rats brazenly moving around during the day, feeding on food in their small backyard gardens, and scurrying about the dumpsters behind neighborhood convenience stores and fast food restaurants.

The massive immigration of individuals into the southern United States from areas in which mosquito-born diseases are endemic as well as deteriorating living conditions due to overpopulation has resulted in increased risk of mosquito-borne diseases. Many of the immigrants come from countries in which dengue and malaria are as common as the cold and flu are in the United States.

Poor exterior sanitation and lack of window screens and air conditioning are the precursors to a major epidemic of mosquito-borne diseases. The yellow fever mosquito breeds in artificial containers and is found throughout the southern tier of states. The recent introduction of the Asian tiger mosquito into this country adds another dengue vector to the arsenal and extends the potential range of the disease further north. Both species thrive in deteriorating conditions where artificial containers are abundant. These mosquitoes readily enter structures and bite humans during the day.

In the 1970's, fear of recurrence of malaria transmission grew when millions of Vietnam veterans, many of whom were infected with the disease, returned to the United States. Despite the presence of a major mosquito malaria vector in the United States a major epidemic did not transpire, due primarily to the prophylaxis used during the war as well as to aggressive treatment of infected individuals.

The recent recurrence of this disease in the United States can likely be attributed to massive immigrations of infected individuals from endemic areas as well as to their proximity to disease vectors.

Stings by envenomizing arthropods such as bees, wasps, scorpions, spiders, and fire ants are increasing because of human encroachment into areas in which these pest species are prevalent as well as to destruction of their native habitats. Increased pest populations and colonizations within structures increase both the risk of human exposure and the possibility of fatal anaphylactic reactions in sensitive individuals.

The northern migration of the aggressive Africanized honey bee and its limited incursion into the United States is another example of how an introduced pest can get out of hand and become a significant threat to human health and safety.

The responsibilities of pest management technicians extends well beyond the exterior walls of structures into the surrounding environment. As a protector of health, the attention of technicians can no longer focus solely on control of cockroaches, fleas, ants, termites, and stored product pests but must also consider the potentially devastating effects these disease vectors can have on customer health.

IMPACT ON PROPERTY

The most significant damage to property within the United States is caused by termites which readily feed on wood in structures and other cellulose-containing materials. The National Pest Management Association (NPMA) reports that termites attack 600,000 homes annually, causing more than $1.5 billion in damage.

Because the damage to structures is usually slow, it typically goes undetected for years so that the cumulative effects can be staggering. Eventually, the termite feeding activity can cause floors and ceilings to sag, walls to crack, and wood to become structurally unsound, resulting in the need for major structural repairs.

Rodents, especially rats, are a close second to termites in causing property damage and are responsible for $500 million to $1 billion in property damage per year. Approximately 20% of fires of unknown origin are suspected as having been caused by rodents gnawing on electrical wiring. They also gnaw holes through walls, doors, and cabinets, etc. in order to gain access to and move throughout the structure in their search for food, water, and nesting sites. They destroy millions of dollars worth of food prod-

The protection of property from wood-destroying pests and organisms is one of industry's major challenges.

ucts and carry several diseases which affect human and domestic animal health.

Numerous pests that cause property (e.g., wood fabricated items, fabrics, leather goods, stored products, etc.) damage include wood-destroying beetles, wood decay fungi, birds, clothes moths and carpet beetles.

IMPACT ON ENVIRONMENT

Some pest management companies appear to be "stuck in a rut"; i.e., they don't know how to incorporate Integrated Pest Management (IPM) or reduced pesticide use strategies into their pest management programs. Many don't perceive the need to change what they are doing; why should they when only a few customers ask for IPM and even fewer know what it means? Finally, some pest control companies are convinced that the only way to make a living is to spray baseboards monthly.

Fortunately, only a few companies within the industry have experienced the unpleasantness and expense of a misapplication lawsuit. Pest management technicians, by serving as guardians of the environment as they utilize pest management methods, play a very important role in limiting liability for pesticide misapplication.

Within the industry, reduction in pesticide use is being achieved in many ways. The era of the monthly service is drawing to a close. More companies are now offering quarterly, semiannual and annual service, thereby reducing the amount of product being applied. Some companies offer non-toxic pest control and have reduced their reliance on pesticide application by increasing the time spent on identification of pest problems and elimination of contributing factors. Preventive measures have resulted in significant reductions in pesticide use indoors with concomitant increased emphasis on applications to the exterior perimeter.

The impact pesticides have on the environment is more closely questioned and observed in today's environmentally-sensitive society.

Many manufacturers are seeking means to reduce risk from the use of their products. Several new baits for use on cockroaches, ants, termites, etc., are now available that incorporate the concepts of pinpoint application, low toxicity, and low dosage. Insect growth regulators, chitin synthesis inhibitors, and other novel technologies targeted specifically for use on arthropods have been developed. New general-use pesticides and termiticides are being applied at remarkably low concentrations, some as low as 0.03%. All of these innovations have resulted in reduced use of pesticides. The industry has come a long way since the 1950's, '60's, and '70's when pesticides were applied indiscriminately. We now have tools at our disposal which enable us to reduce risk and use of pesticides, satisfy our customers, and achieve our pest management goals.

CONCLUSION

The staggering impact of pests on food, health, property, and environment costs consumers billions of dollars annually. Pest management technicians are among those with the important responsibility for minimizing the effects of pests and their economic impact while, at the same time, protecting the environment.

CHAPTER 2

Integrated Pest Management

History and Definition
Procedures
Techniques

Pest management has changed significantly throughout its history. At the turn of the century, there were very few pesticides, most contained lead, arsenic, nicotine, and a variety of other acutely toxic materials, however, the industry relied heavily on the use of nonchemical strategies to manage pests. During and after World War II, pesticide development and production increased significantly and an era of almost exclusive pesticide use began in the industry. Then in the 1980s, reliance on pesticides began to decrease, and the industry began moving toward reduced-risk pest management strategies, i.e., less pesticides and more nonchemical procedures. Thus integrated pest management (IPM), a well established agricultural term, was adopted by the structural pest management industry.

WHAT IS IPM?

IPM originated in agriculture and is based on the concept that there are agricultural pest levels which can be tolerated without suffering significant crop production losses and the cost of pesticide use is not justified. However, if the pest population exceeds a predetermined threshold, insecticides are applied to lower the population below the economic threshold.

The agricultural concept of IPM poses some problems for the structural pest control industry. Economic thresholds within the structural pest management industry are not commonly used because the customers' aesthetic (economic) threshold is zero and their expectations often mandate that the pest problem be resolved immediately.

Occasionally, action thresholds have been used within structural pest management, for instance the Department of Defense in a recent publication suggests that an average sticky trap catch of fewer than one cockroach per trap, per night does not justify pesticide application. It further suggests that trap catches of one to three cockroaches per trap, per night may require a spot treatment. However, this is an exceptional case in structural pest management.

The National Pest Management Association (NPMA) and United States Environmental Protection Agency (USEPA) have formalized two definitions which better describe the concept of IPM within the structural pest management industry. NPMA defines IPM as a decision-making process that anticipates and prevents pest activity and infestation by combining several strategies to achieve long-term solutions. Components of an IPM program may include proper waste management, structural repair, maintenance, biological and mechanical control techniques, and pesticide application.

The USEPA in their IPM in schools brochure defines IPM as an effective and environmentally sensitive approach to pest management that relies on a combination of common sense practices. IPM programs take advantage of all pest management options, possibly including, but not limited to, the judicious use of pesticides.

A few organizations characterize IPM as an approach to pest management which considers the use of pesticides as an independent entity and not as a component of the program. This flies in the face of what IPM is and what its goal should be — a decision-making process and planning program designed to identify specific pests and infestation site(s), suppress the infestation with short-term solutions, and reduce the causes of infestation with long-term strategies. IPM is proactive rather than reactive.

Good sanitation is essential to an IPM program.

An IPM program can include any or all of the following elements:

• Cultural practices are those factors influenced by humans and their environments. Sanitation is the most important cultural practice affecting the success of a pest management plan and it is a customer responsibility. Sanitation removes the essential elements for pest survival, i.e., food, water, and harborage. Specific examples are waste removal, cleaning, grass mowing, exterior clean-up, etc.

• Biological techniques involve the use of living organisms or their by-products to control pests. Parasites and predators are very useful in controlling outdoor pests, however, there is very little application indoors. Fungi, bacteria, and other microorganisms have been used successfully in a few situations indoors, and one of the most popular baits in the industry is derived from a bacteria. Insect growth regulators and chitin synthesis inhibitors, which abnormally affect growth and development, are included in this category.

• Physical techniques typically involve the use of heat or cold to control pest populations. Depending on the exposure time, temperatures above 140⁰ F are usually sufficient to kill most arthropods. Cold temperatures can be used to kill most stages of arthropods, however, temperature and time of exposure are more variable than when using heat.

Biological control products use living organisms, like this wasp, or their by-products to manage pests.

• Mechanical devices have for hundreds of years been used to manage pest populations. There is probably no household in the United States which does not have a fly swatter. There are wind-up live traps, sticky traps, snap traps and a variety of other devices for controlling rodents.

Pesticides kill or manage insects by affecting their behavior or growth.

Screening, netting, hardware cloth, caulking, and expandable foam, are examples of just a few of the products available for excluding pests from structures. These tools are very valuable in preventive pest management.

• Pesticides are products designed to kill or manage pests by affecting their behavior or growth, e.g., repellents, insect growth regulators, plant growth retardants, etc., and they are a very important component of /IPM/. The selection of a particular product depends on the site, pest, active ingredient, and the formulation.

The phrases "innovative pest management" or "creative pest management" better describe the type of service the industry should be providing. This terminology conveys the concept of thinking before reacting and is founded on the principle that each pest management situation is different and has its unique solution. There are no quick fixes in this industry, if there were cockroaches would have been eradicated a long time ago. One of the industry's leading educators, Dr. Austin Frishman, says, "Technicians need to think like a roach." Successful pest management programs depend on locating problem areas and harborages that are not always apparent. Technicians play a pivotal role in the success of IPM programs. Their specific responsibilities can best be characterized as education, investigation, customer education, implementation, and observation.

ROLE OF EDUCATION

As pest management professionals, technicians must not only be skilled in pest identification but must be familiar with pest biology, pest behavior, pest management strategies, and pesticides before attempting to solve their customers' pest problems. They must be sufficiently knowledgeable to explain why and how they plan to deal with specific pest management situations.

There are numerous educational opportunities within the pest management industry. Formal training is provided through several universities, community colleges, and technical schools. Many companies have in-house training programs which utilize a variety of training materials, e.g., computer training programs, slide tape series, videotapes, workbooks, etc. Educational programs are offered at NPCA and state association meetings and by most product manufacturers. Self education is possible by reading and studying any of the numerous industry reference materials.

To educate the many customers served by this industry technicians must first be educated themselves. Pest management technicians should consider themselves the hub of a wheel with information radiating out from the center.

PERFORMING THE INSPECTION

This is the fundamental step in innovative pest management and is critical in minimizing pesticide applications and providing long-term solutions. Solving pest management problems depends on the synergy between inspection and identification, because one is worthless without the other. The benefits of a good inspection and correct identification, are knowledge of the pest, its biology and behavior, and solution of the pest problem.

The inspection is your opportunity to become Sherlock Holmes, and to solve an unusual pest management problem and succeed where others have failed. It is very gratifying to solve a tough "bug" problem and see the reaction of satisfied customers when they are informed their home or business is pest free. The success or failure of the pest management plan is determined by the thoroughness and accuracy of the inspection, and this requires time. Some problems are easily diagnosed, for instance walking into a structure and seeing German cockroaches climbing the walls during the day or fleas jumping onto pant legs, however, recognition of the pest isn't enough.

The inspection should begin with a detailed customer interview in order to determine customer expectations and concerns, but most importantly the type and extent of the pest problem. This information-gathering session should provide some clues as to whether or not there is an actual pest problem. The worst thing a technician can do is assume that the customer has identified the pest problem correctly and base the pest management plan on this assumption.

This error in judgment has led many technicians down the primrose path of treating for imaginary pests such as the "cable mite" or "paper mite." Misidentifications of this type are not only embarrassing for the technician and their company, but is definitely not the type of service upon which good reputations are built. These situations occur when technicians fail to properly inspect and identify the problem, and should never occur if the proper inspection techniques are followed.

CUSTOMER EDUCATION

Once the pest problem is identified and a pest management plan has been developed, customer education becomes a priority. This phase of the pest management plan involves explaining procedures and products to be used, determining the customer's expectations, and explaining their role in the resolution of their pest management problem. If this is not accomplished, customers may have unreasonable expectations.

Unfortunately most consumers have no idea what IPM is, therefore, technicians have a formidable task ahead in terms of educating consumers about this type of service. IPM relies very heavily on the use of nonchemical pest management strategies and many pest problems are corrected through structural changes and/or repairs, which typically are the customer's responsibility.

While it is common for pest management companies to caulk and/or stuff cracks and crevices, install door sweeps and other exclusion materials, major structural problems, e.g., roof leaks, slab cracks, moisture problems, lighting, etc., are beyond their capabilities and require customer resolution. Typically, if these major structural problems are not resolved the pest problem persists. Education is an ongoing task, and everyone in the industry must play an active role.

IMPLEMENTING SERVICE

This is the step in innovative pest management when knowledge and experience are put into practice by implementing the pest management plan presented to the customer. This phase of the IPM process is continuously changing and requires periodic review with the customer. The review not only ensures the success of the plan, but customer satisfaction. Control failures rarely occur because of pest resistance to pesticides, they typically occur because the pest problem was not adequately evaluated, lack of knowledge about the pest's biology and behavior, and/or failure to use pesticides in appropriate locations.

The industry has an emerging arsenal of pesticides, such as insect growth regulators (IGRs), chitin synthesis inhibitors, baits, biologicals such as nematodes, bacteria and fungi, and other new active ingredients. To supplement these pesticides there are improved delivery systems, such as aerosol units, bait stations, power dusters, and other tools to accomplish more precise applications. After determining the type of pests and their location, the proliferation of products and tools in the industry presents the service technician with an overwhelming number of decisions, such as selecting the best products and/or techniques to use, addressing the customer's concerns about pest management services, time of service, etc. These decisions influence the implementation of the pest management plan.

Time is a major factor when implementing a pest management plan and frequently determines its success or failure. Time influences the essential ingredient of any plan, which is thoroughness in execution. This alone reduces costly and time-consuming callbacks and has an immediate effect on customer satisfaction. The technician's time must be delicately balanced in order to meet the customer's needs and their company's performance objectives.

The implementation phase of the IPM program is where the leather hits the road and to be successful technicians must consider all of the factors affecting the pest management plan and be innovative in their approach to the pest management problem.

ACCOUNT OBSERVATION

The success of the pest management plan is predicated on the technician's knowledge of pest management, a thorough inspection which properly identifies the pests and contributing factors, educating the customer and obtaining their cooperation, and full implementation of the IPM plan.

Each component determines, to some degree, whether the customer's pest problem is successfully resolved. Evaluating the success of the pest management plan is the final step in the IPM program. Follow-through is the path to success in managing any pest problem and an opportunity to learn and improve pest management skills.

The pest management plan can have several outcomes, e.g., elimination of the pest problem, no apparent change in the pest problem, accusations of feeding the cockroaches, etc. However, it is more likely that some degree of success between these extremes has been achieved. This underscores the importance of continuing to observe and evaluate the account until the problem is satisfactorily resolved.

One of the best methods of resolving problems is to have a documented history of the account, e.g., diagrams, information on previous pest problems, locations of infestations, techniques employed, etc. The more detailed the record the better.

Records provide a starting point for problem resolution. Systematically reviewing the records and merging that information with correct observations can lead to a timely solution for many problems. This helps to identify continuing pest problems by observing activity in different sites, discovering previously undetected sanitation problems, harborage or pest entry points, uncovering new pest problems, etc. Logbooks are typically used to retain this information and track the pest management program in an account or facility.

German cockroach and termite control failures are the most common in the industry, and there are a host of factors which can cause these control failures. One of the most common reasons cited for cockroach control failures is resistance. While this is occasionally a problem in some commercial accounts, it can be managed with a thorough inspection, and appropriate selection and use of pesticides.

Customer involvement, particularly regarding sanitation, proper storage and structural repairs is often a major obstacle. Failure to identify all sites of pest activity and harborage areas which require treatment also leads to control failures. Improper pesticide selection and failure to thoroughly apply the product to pest harborages and activity sites can defeat the best pest management plan. The importation of cockroaches by vendors and their immigration from adjacent facilities must also be considered.

Termites pose completely different challenges for technicians because they are usually hidden and in this instance technicians aren't, typically, engaged in the process of pest elimination, but are simply attempting to exclude them from the structure and their food source. Careful observation on callbacks is critical if the problem is to be resolved. Identifying the species is a critical step because it can not be assumed that native subterranean termites are the problem. Occasionally, Formosan, drywood, and dampwood termites are encountered and each has unique biological characteristics and habits which require special control techniques.

Once the species is identified, it is extremely important to determine where the pest activity is continuing and assessing the potential causes. This, of course, assumes everything has been done to prevent the infestation, e.g., using the maximum label concentration and rate of application practical, mixing the termiticide adequately, calibrating equipment for accurate delivery rates, drilling holes at the proper locations and distances, properly treating porches and chimney bases, etc.

Often overlooked is the presence of structural problems, e.g., a moisture problem in the structure that allows a native subterranean termite colony to become established in the structure out of contact with the soil, wood soil contact, inaccessible crawlspaces, foam insulation board in contact with the soil, which can provide an undetected conduit between soil and the structure, etc.

There are many other factors to consider with regard to the pests listed previously, as well as many other pest problems. The technician's knowledge of pests is the key element in analyzing observations and taking the necessary corrective action. Spending sufficient time making thorough observations pays its dividends by reducing the number of callbacks and insuring customer satisfaction.

CHAPTER **2**

CHAPTER 3
Inspection and Monitoring
Where to Look
Tools and Techniques
Collecting/Preserving Specimens

The most important objective of the inspection is to determine if there is a pest problem, and, when one is found, the solution will depend on the synergy between inspection and identification; one is worthless without the other. A good inspection and correct identification result in knowledge of the pest, its biology and behavior, and solution of the pest problem. Monitoring of pest populations, the logical extension of inspection procedures, is then provided in order to determine areas of pest activity, population size, and the success of pest management procedures.

INSPECTION TOOLS

Sticky traps are designed to trap crawling insects, such as cockroaches and occasional invaders, and flying insects, such as flies and stored product pests. Most traps are composed of paper material which can be folded into a tent or box for crawl-

Sticky traps help technicians identify insect activity.

ing pests or they are large plastic flat boards, tents or tubes for flying insects with one characteristic in common — at least one surface is coated with an adhesive material.

Some traps are baited with food attractants or contain a pheromone (a chemical used by insects to communicate among species). Traps usually are used in situations where monitoring is conducted over an extended period of time. When they are reused, traps should be dated at the time of placement and every time they are inspected. Other recordings should include the type and number of pests. Traps should be replaced when they become damaged, wet, or full of insects.

Lights are used to conduct visual inspections. For ease of handling and maximum brightness, flashlights should be small and have high intensity bulbs. Rechargeable flashlights are usually brighter and more compact but can be used only for short periods of time before they must be recharged. Replaceable battery flashlights are bulkier and do not have the intensity of most rechargeable lights, but they do not require recharging. Extra batteries should always be available.

Flushing agents, e.g., pyrethrin aerosol, heat, and air can be used to stimulate cockroaches and other insects to exit deep harborages. These products, which require several minutes to become effective, are of greatest value when there is evidence of insect activity, such as droppings, specks, and body parts, despite the lack of readily visible live insects.

A good flashlight is essential to conduct a proper inspection.

Mirrors are essential tools for use in inspection of commercial kitchens and other areas in which line of sight inspections would be difficult or impossible due to the amount of equipment. The best mirror is one that adjusts on the end of a telescoping rod in order to facilitate inspection of areas deep in cabinets and up and behind pipes, splash boards, sinks, etc.

Spatulas can be used to remove debris that is obstructing the inspection, to reach deeper into cracks and crevices where insects are often breeding out of sight, and to collect specimens which then can be placed in containers for further identification.

Hand lenses, which allow one to make field identifications, should have at least 10X magnification power. Magnification devices which have 35X power are available but are cumbersome.

Protective equipment varies depending on the account being serviced, the area within the account to be inspected, and the suspected pest problem present in the account. Kneepads, coveralls, boots, bump cap, respirator, goggles, and gloves should be readily available for use by technicians if necessary. For example, kneepads and light-weight rubber gloves are most commonly used during cockroach inspections; coveralls, leather gloves, a bump cap, and goggles or safety glasses are commonly used during termite inspections in a crawlspace.

Specimen containers are used to return specimens to the office for further identification. Pill boxes, prescription bottles, and plastic specimen containers with snap lids typically are used to contain specimens.

Miscellaneous inspection equipment includes ladders, termite detection equipment, e.g., probes for sounding wood, moisture meters for detecting above-ground activity, methane gas detectors, electronic sound detection devices, a stethoscope, and a lighted scope for examining wall voids, and a UV-light for detecting rodent urine.

INSPECTION PROCEDURES

Successful solutions to difficult pest management problems often require changes in human habits. Tough problems might necessitate performing an inspection at night or crawling into a tight crawl space or hot attic. Unwillingness to perform these less-than-desirable tasks quite often leads to the failure of the pest management plan and to the continuation of the pest problem.

Success within the pest management profession cannot be attained without the ability to crawl, kneel, climb, and perform numerous other bodily contortions in order to conduct inspections and achieve "eye-to-eye" contact with structural pests. Dr. Austin Frishman, a noted industry consultant, considers kneepads to be the single most important piece of equipment that a technician can take into any account because they enable the technician to get down to pest level.

Another aspect of the inspection is timing. The activity patterns of most pests do not always coincide with that of humans. Many pest species are active at night, a time when they are least likely to encounter humans and other potential threats to their survival. For instance, most cockroach species are active at night, readily moving in response to any disturbance, e.g., air, vibration, and light. Carpenter ants are active at night and are rarely encountered during the day except in small numbers. Termites, which are very susceptible to environmental conditions, are rarely seen, and their colonies are never seen as they remain protected out of sight within the soil.

A comprehensive inspection to identify existing pest infestations and factors that may contribute to future problems should be performed on every new account, i.e., the structure should be inspected inside, outside, over and underneath, if applicable. In order to conduct a thorough inspection, some knowledge of common pest biology and habits is essential. This knowledge is also valuable when a pest problem develops in established accounts because it focuses the inspection on likely areas of pest harborage and activity.

In order to minimize the chance of missing a problem area, the inspection should be conducted systematically through and around the structure. Documentation of the inspection, e.g., floor plans, checklists, and notes, is useful in analyzing problem(s) when they develop and monitoring the progress of the pest management program.

PEST IDENTIFICATION TIPS

Pest identification is one of the most important steps of the inspection process. If a pest can not be identified from experience or from some unique characteristic, e.g., unusual form, color, or behavior, magnification will facilitate identification through observation of other physically-distinctive characteristics. The process can be facilitated further by consulting Chapter 4 of this Handbook as well as reference specimens, taxonomic keys found in texts such as the recently-published *PCT Field Guide* series and the *NPMA Field Guide to Structural Pests*, and/or a technical expert in your company, consultant, the NPMA, and/or state extension service.

Insect identification is an important task for technicians.

When necessary, specimens should be carefully collected and transported back to the office for identification. The better-protected a specimen is during collection and transport, the easier it will be to identify. There is no joy in trying to piece together an arthropod for identification.

Live and recently-killed specimens should be placed in a container which will not be affected by moisture; if the specimen will be stored for more than a day, it should be placed in a vial containing sufficient alcohol (70% ethanol is best) to cover the specimen and then tightly sealed. Dead specimens become extremely dry and brittle, and, therefore, must be handled with care to avoid breaking off body parts or completely crushing and destroying the specimen. Dry specimens should be placed on layers of tissue paper (i.e., not cotton or other fibrous materials) within a container with additional paper loosely- placed over the specimen to cushion it. Sticky boards, scotch tape and other adhesives should not be used to collect specimens because once the specimen is stuck in the glue, identification is much more difficult.

Once correct identification is made, information regarding the pest's biology and behavior can be obtained by consulting various resource materials available through trade magazines, national and state pest control associations, extension offices, distributors, etc. This information, in addition to the on-site inspection report, will be used to develop an effective and appropriate pest management plan that takes into consideration the pest, its abundance, biology, behavior, active sites, and customer needs and expectations.

PEST MONITORING

Monitoring detects emerging pest problems and evaluates the success of the pest management program. Monitoring is typically conducted by using sticky traps, visual inspections, and/or customer sightings. Each of these methods has advantages and disadvantages.

Successful monitoring with sticky traps depends on placement and knowledge of the pests' biology and habits. It is essential that traps be placed in the path of the pests or close to active areas. When traps are used correctly, they provide a visible record of the success of the pest management program and provide ability to perform continuous monitoring over an extended period of time. They should be inspected every two to four weeks, and customers should be encouraged not to remove or damage them.

Visual inspections, which will be aided by the use of flushing agents, rely even more heavily on knowledge of pest biology and habits, as well as ability to recognize the signs of infestation. The major limitations to visual inspections are the length of time required in order to perform the inspection and the activity patterns of the pests (which, typically, are more visible at night). Advantages include a better working knowledge of the account, identification of contributing factors to pest problems, better knowledge of active areas, and increased interactions with management through follow-up discussion. Customer sighting, although a useful tool, is the least desirable method of monitoring pest activity.

A thorough inspection and subsequent monitoring are critical steps in the development and execution of a successful IPM plan. It is likely that, in the future, technicians will perform more monthly inspections and monitoring and less monthly spraying. This trend is currently evident in the numerous IPM programs that have been implemented in facilities such as schools, hospitals, and government buildings.

CHAPTER 4
Identification and Control
Taxonomic Basics
Insect Development and Life Stages
Key to Pests in this Book
Pest Groups

As previously stated in Chapter 3, "A good inspection and correct identification result in knowledge of the pest, its biology and behavior, and solution of the pest problem." This chapter is designed to assist in the identification of common structural pests and to provide information on their basic biology, habits and control.

TAXONOMIC CLASSIFICATION

All living things are classified using a binomial system, i.e., they are named according to their genus and species. The method of classifying plants and animals is described below with the divisions listed in order of increasing specificity.

Kingdom (Animal) — This major division includes most living organisms which are capable of motion and do not have chlorophyl in contrast to the plant kingdom which includes most living organisms which are immobile and contain chlorophyl.

Phylum (Arthropoda) — This is the first major taxonomic unit that includes animals which share fundamental patterns, organization, and descent. Arthropods are characterized as having jointed appendages, an exoskeleton, dorsal heart, ventral nervous system, and cold blood.

Class — There are many classes in each phylum; these classes are further subdivided into orders. The five major arthropod classes are arachnida, chilopoda, crustacea, diplopoda, and hexapoda.

Order — Classes are subdivided into orders–groups of living things with major common characteristics such as number and types of wings, types of legs and mouthparts, and type of development.

Family — Orders are subdivided into families which describe groups of animals with a wider range of common characteristics. Most family names are recognized by their suffix "-idae." The ability to identify a pest to family often provides sufficient information so that a pest management program can be implemented.

Genus — Once identification reaches this point, sufficient information exists on biology, habits, and control for the problem to be effectively managed. The name of the genus is always capitalized and italicized or underlined.

Species — The last major subdivision in the classification scheme, species describes a distinct group of life forms which has well-defined characteristics in common and is capable of producing offspring with the same characteristics. It is helpful to be able to identify pests to species because it provides more specific information on their biology, habits, and control. The name of the species is always in lower case and italicized or underlined.

The following are examples of how two of the most common structural pests are classified:

Common Name	German Cockroach	House Mouse
Kingdom	Animal	Animal
Phylum	Arthropoda	Chordata
Class	Hexapoda	Mammalia
Order	Blattodea	Rodentia
Family	Blattellidae	Muridae
Genus	Blattella	Mus
Species	germanica	musculus

When reference is made to a species, both the genus and species is given, e.g., German cockroach (*Blattella germanica*) and house mouse (*Mus musculus*).

INSECT GROWTH AND DEVELOPMENT

Insects hatch from eggs and progress through several stages (instars) before reaching the adult and reproductive stage. The growth of the insect is limited by the size of the exoskeleton; in order to grow, the exoskeleton must split, thus, allowing the more mature developmental stage to crawl out and form a larger exoskeleton.

During this process, the immature insect appears white (this often generates reports of albino insects), but as the exoskeleton hardens, it becomes darker. Once an insect reaches the adult stage, it is unable to grow any larger. Thus, there are no such things as "baby" flies; the flies are simply smaller due to environmental and food conditions during their growth and development process.

Metamorphosis is the process by which insects grow and change in form from the egg to the adult. As they progress through these developmental stages, they molt several-to-many times depending on the species; each molt results in an instar. The instars are numbered according to the number of molts through which the immature insect has gone. The development time from egg to adult can be as short as a week to as long as seventeen years for the periodical cicada. The two common types of metamorphosis are simple and complete.

Simple metamorphosis involves a gradual change, i.e., the immature insect looks more and more like the adult with each successive molt. In insects with wings, external wing pads develop during the early instars; compound eyes, if present in the adult, are developed in the early instars, and there is no inactive stage prior to adult emergence.

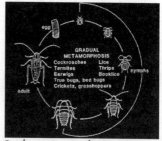

Simple insect metamorphosis.

The immature insects are called nymphs (i.e., terrestrial species) and naiads (i.e., aquatic species). Pest management of insects with this type of development is simpler because the nymphs and adults have similar appearance, biologies, and habits; therefore, the control strategies are basically the same. Common structural pests which undergo simple metamorphosis include cockroaches, crickets, lice, earwigs, and sowbugs.

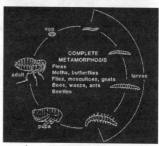

Complete insect metamorphosis.

Complete metamorphosis involves four developmental stages — egg, larva, pupa and adult, none of which resemble each other during development. In this type of development, the wings develop internally, compound eyes are not present in the larval stage, and there is a prolonged of inactivity (pupal stage) prior to adult emergence. Insects which undergo complete metamorphosis pose a greater pest management challenge because the larvae and adults do not look alike, they have different biological characteristics and habits, they often cause different types of damage, and the pupal stage is resistant to most pest management strategies. Thus, these species of insects are generally more difficult to control and may require several pest management strategies. Common structural pests undergoing complete metamorphosis include ants, bees, wasps, flies, fleas, beetles, and moths.

INSECT STRUCTURE

This section describes the major anatomical features of the three insect body regions (head, thorax, and abdomen) which are often used for identification.

An insect is made up of three primary body regions: head, thorax and abdomen.

Insect Head

Antennae — The antennae are the major sense (i.e., touch, smell, and hearing) organs found on the head. They are of various forms and shapes, e.g., filiform — thread-like (e.g., cockroach), moniiliform — bead-like (e.g., some bark beetles), serrate — saw-like

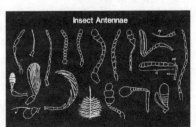

An insect's antennae are its major sense organs for touching, smelling, and hearing.

(e.g., drugstore beetle), clavate — clubbed (e.g., ladybird beetle), and

capitate — having a head (e.g., some powderpost beetles).

Eyes — Most insects have a pair of large compound eyes composed of many small lenses. Some insects have one-to-three **ocelli**, i.e., simple eyes, located between the compound eyes.

Mouthparts — The mouthparts of most insects are located on the lower part of the head; they project downward, although there are a few species in which the mouthparts project forward. Mouthparts are very useful in insect identification and provide some insight into the type of food eaten by the insect. Six types of mouthparts are commonly described for insects:

Chewing mouthparts are the most primitive and basic mouthparts found in insects, e.g., cockroaches, earwigs, crickets, many larvae, beetles, and termites. They are composed of seven segments, two mandibles (i.e., jaws), an upper lip (i.e., labrum), a tongue (i.e., hypopharynx), two lower jaws (i.e., maxillae), and a lower lip (i.e., labium). All other types of mouthparts are composed of the same components although they are often modified significantly.

Rasping-sucking mouthparts are found in thrips which use them to tear the surface of plant tissues, thus, causing fluids to flow out of the plant.

Piercing-sucking mouthparts are usually long and slender and needle-like. They are found in a variety of insects, e.g., mosquitoes, biting flies, fleas, lice, predators, and plant bugs with significant modifications among species.

Sponging mouthparts are found in filth flies and fruit flies; they are adapted for sucking up liquefied food. The base of the elongated mouthpart is shaped into a large fleshy sponge which has a series of grooves radiating from the central food channel.

Siphoning mouthparts are found in butterflies and moths. The seven elongated mouthpart components are fused to form a long siphoning tube which is used for sucking up plant nectar and water. These mouthparts are coiled up at the base of the head at rest.

Chewing-lapping mouthparts are found in some bees and wasps. The structure of these mouthparts allows these insects to chew solids and suck up exposed liquids.

Immature insects undergoing simple metamorphosis typically have the same mouthparts as the adults. However, there is more variability in the mouthparts of larvae undergoing complete metamorphosis. For example, moth and butterfly larvae have chewing mouthparts; most beetle adults and larvae have chewing mouthparts; and ant larvae have reduced chewing mouthparts.

The Insect Thorax

The thorax, the middle section of the insect body, is composed of three segments which, listed from front to rear, are the prothorax, mesothorax, and the metathorax.

Legs — Insect legs are always jointed; one pair is attached to each

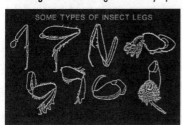

An insect's legs are always jointed and one pair is attached to each thoracic segment.

thoracic segment. The leg has six parts — from their attachment to the thorax outward, they are the coxa, trochanter, femur, tibia, tarsus, and pretarsus. The tarsus is composed of several segments; the pretarsus typically has a pad and one or two claws. legs come in all shapes and sizes and are modified for a variety of activities, e.g., running, jumping, swimming, digging, and grasping. Legs are used for identification and provide insight into the insect's biology and habits.

Wings — Most insects have two pairs of wings which are outgrowths of the last two thoracic segments. However, some insect species have only one pair of wings; a few have none. On flies, the second pair of wings is modified into balancing organs called halteres. Most insect wings have a cellophane-like appearance and few-to-many veins which often are used for identification. The first pair of wings on most beetles is modified into hard wing coverings called elytra. In some insects, either a portion of or the entire first pair of wings is leathery. Variation in the number of wings, size, position, veination, and shape are clues to identification.

The Insect Abdomen

This is the third body region which has numerous segments, the spiracles and external reproductive organs. Spiracles are the exterior opening of the respiratory system; some insects have one pair on each abdominal segment. In most insects, the external reproductive organs used for copulation are on the last two abdominal segments. Some insects, such as cockroaches, crickets, and silverfish have segmented appendages called cerci which extend from the tip of the abdomen.

Class Arachnida. The characteristics of this class include: no wings and antennae, one or two body segments (i.e., head/thorax and abdomen), and four pairs of walking legs. There are terrestrial and freshwater species. Major orders include:

Acari (mites and ticks) — The head, thorax and abdomen appear to be combined, however, they have the two body regions typical of arachnids — cephalothorax and abdomen — and often are sac-like in form. The adults and nymphs have four pairs of legs; larvae, however, have only three pairs. They have piercing sucking mouthparts, and the central tube (hypostome) in ticks has recurved teeth; mites, however, lack teeth. Ticks and mites undergo simple metamorphosis (i.e., egg, larva, nymph and adult).

Araneae (spiders) — Spiders have two body regions — a cephalothorax (head and thorax) and an abdomen — which are connected by a tiny waist (pedicel). The abdomen is bulbous and unsegmented and has several silk producing spinnerets at the tip. Spiders have four pairs of legs, and most species have six or eight simple eyes. A pair of jaws which end in a hollow fang is located below the eyes. Most spiders have venom although most are incapable of biting. Spiders undergo simple metamorphosis (i.e., eggs, spiderlings, and adults).

Scorpiones (scorpions) — These arachnids have two body regions—cephlaothorax and abdomen. The long segmented abdomen usually is curved upward ending in a stinger which contains a venom gland. A large pair of pincers are attached to the cephalothorax and are used to capture and hold prey.

Class Crustacea (crabs, sowbugs, and copepods). The characteristics of this class are as follows: wingless, two pairs of antennae, two body regions (i.e., cephalothorax and abdomen), and ten or more pairs of legs. Most species are marine, but there are a few freshwater and terrestrial species.

Isopoda (pillbugs and sowbugs) — These isopods are oval-shaped and appear to have body armor. They do not have a cover over the cephalothorax (i.e., head and thorax combined). They have two pairs of antennae, the second of which is very small, and seven pairs of legs.

Class Chilopoda (centipedes). The characteristics of this class include: wingless, one pair of antennae, flat worm-like body with many segments, and a single pair of legs on most body segments. All species are terrestrial.

Class Diplopoda (millipedes). The characteristics of this class include: wingless, one pair of antennae, cylindrical worm-like body with many segments, and two pair of legs on most body segments. All species are terrestrial.

Class Hexapoda (insects). The characteristics of this class include: one pair of antennae, three body regions (i.e., head, thorax and abdomen), three pairs of legs, and one or two pairs of wings which sometimes are absent. Most species are either terrestrial or freshwater, but a few are marine. Major orders of structural pests include:

Anoplura (sucking lice) — These are small, flattened, and wingless blood-sucking parasites of warm blooded animals. The mouthparts, which are adapted for piercing and sucking, are withdrawn into the head when the insect is at rest. The compound eyes are either small or absent. The tarsus is single-segmented with one large claw adapted for hanging onto hairs. Lice undergo simple metamorphosis (i.e., egg, nymph, and adult).

Blattodea (cockroaches) — These insects are easily recognized by their oval-shaped and dorsally-ventrally flattened body. The head is concealed from view by the first dorsal thoracic shield (i.e, pronotum). They have four wings; the first pair has veins and are parchment-like and completely cover the second pair which are cellophane-like. A few species in this class of insects lack fully developed wings and have short stubbly wing pads. The antennae are long and thread-like, and the mouthparts are chewing. The tip of the abdomen has two segmented projections (i.e., cerci), however, they are not pincer-like. Cockroaches undergo simple metamorphosis (i.e., egg, nymph, and adult).

Coleoptera (beetles and weevils) — In most species, the front pair of wings is thick and shell-like, meet in a straight line down the middle, and serve as wing covers (i.e., elytra) for the second pair of cellophane-like wings folded underneath. Antennae usually have eleven or fewer segments and range in size from inconspicuous to more than twice the body's length. The mouthparts are chewing and, in the weevils, are at the end of a snout or projection from the front of the head. Beetles and weevils undergo complete metamorphosis (i.e., egg, larva, pupa, and adult).

Collembola (springtails) — These very small insects are wingless, and their abdomens have six or less segments. Most species have a forked structure (i.e., furcula), the "spring" attached to the tip of the abdomen normally tucked under the body. In addition, there is a tubular structure (i.e., collophore) on the underside of the first abdominal segment. They have short antennae and chewing mouthparts. Springtails undergo simple metamorphosis (i.e, egg, nymph, and adult).

Dermaptera (earwigs) — These are elongate and flattened insects with four pairs of wings, the first of which are short and hardened and the second of which are cellophane-like and folded under the first pair. The antennae are thread-like and about half as long as the body. They have chewing mouthparts and compound eyes. The tip of the abdomen has two segmented pincer-like projections (i.e., cerci). Earwigs undergo simple metamorphosis (i.e., egg, nymph, and adult).

Diptera (flies and mosquitoes) — These insects have one pair of cellophane-like wings, the second pair of wings is modified into knob-like structures (i.e., halteres) used for balance. In contrast to other flies, mosquito wings are covered with small scales. The compound eyes are large, and the antennae are often small and inconspicuous. In mosquitoes and other blood feeding flies the mouthparts are piercing and sucking; in filth flies, the mouthparts are sponging. Flies undergo complete metamorphosis (i.e., egg, larva or maggot, pupa, and adult).

Heteroptera (true bugs) — The first pair of wings is thickened at the base and appear leathery; the remainder of the wing is cellophane-like. The second pair of wings is cellophane-like and concealed by the first pair which are held flat over the body. A few bugs, e.g., bedbugs and bat bugs, are wingless. The mouthparts are piercing and sucking and arise from the front part of the head; however, at rest, the mouthparts are held under the body. Bugs undergo simple metamorphosis (i.e., egg, nymph, and adult).

Hymenoptera (bees, wasps and ants) — These insects have two pairs of cellophane-like wings, the second pair is slightly smaller than the first. Worker ants are wingless; however, many species produce winged swarmers at least once annually. The constriction between the thorax and the abdomen is more noticeable in wasps and ants. In ants, this constriction is referred to as the *petiole* and consists of one or two segments (i.e., nodes). Most species have well-developed compound eyes, the antennae are usually 12 to 13 segments (and elbowed in ants), and the mouthparts are either chewing or chewing-lapping. Bees, wasps, and ants undergo complete metamorphosis (i.e., egg, larva, pupa, and adult).

Isoptera (termites) — These wood-attacking insects are divided into several castes (i.e., divisions of labor) — workers, soldiers, kings, and queens. Once a colony is established, all the individuals are wingless; however, during the year, winged swarmers (i.e., kings and queens) are produced. The two pair of wings are cellophane-like, longer than the body, and equal in size.

Although the workers are commonly referred to as white ants, they lack the narrow waist (i.e., petiole) characteristic of ants. Workers and soldiers do not have compound eyes, the mouthparts are chewing, and the antennae are long and bead-like. Termites undergo simple metamorphosis (i.e., egg, nymph, and adult).

Lepidoptera (butterflies and moths) — Most of these insects have two pair of cellophane-like wings, the second of which is smaller than the first, which are covered with colorful scales that easily rub off when handled. The mouthparts are siphoning and coiled into a tube under the head. The antennae are usually long and slender, although the butterfly antennae end in a knob. Butterflies and moths undergo complete metamorphosis (i.e., egg, larva or caterpillar, pupa, and adult). The caterpillar has chewing mouthparts; most have small legs (i.e., prolegs) with small hooks (i.e., crochets) on the bottom.

Orthoptera (crickets) — The adults are either wingless or have two pair of wings. The larger first pair of wings is long, narrow, and thickened, and the hind wing is membranous and kept under the front pair. They have chewing mouthparts, and the antennae usually are long and have numerous segments. The tip of the abdomen has two segmented projections (i.e., cerci) which are either short and pincer-like or long and feeler-like. Some female crickets have a long projection (i.e., *ovipositor*) extending from the tip of the abdomen which is used to lay eggs. Crickets undergo simple metamorphosis (i.e., egg, nymph, and adult).

Psocoptera (psocids and booklice) — These are very small soft-bodied insects which may or may not have wings. If wings are present, there are two pair which are held roof-like over the abdomen when the psocid is resting. They have chewing mouthparts and long slender antennae. Psocids undergo simple metamorphosis (i.e., egg, nymph, and adult).

Siphonaptera (fleas) — These bloodsucking parasites of mammals and birds are very small, wingless, and have laterally flattened bodies. Their legs are modified for jumping, and they have various patterns of spines on their bodies which assist them in hanging onto animal hairs. The mouthparts are piercing and sucking, the antennae are short and inconspicuous, and the compound eyes may or may not be present. Fleas undergo complete metamorphosis (i.e., egg, larva, pupa, and adult).

Thysanura (silverfish and firebrats) — These tear drop-shaped, wingless insects have scales on their bodies, chewing mouthparts, and three long tail-like appendages at the tip of the abdomen. Silverfish and firebrats undergo simple metamorphosis (i.e., egg, nymph, and adult).

The following key to the arthropods of public health importance is included for assistance in identifying structural pests.

Key To Adult Arthropods Described In This Book

1. With three or four pairs of legs 2

    With five or more pairs of legs 3

2. With four pairs of legs; no antennae — spiders, ticks, mites, and scorpions (Class Arachnida) 5

    With three pairs of legs and one pair of antennae — insects (Class Insecta) 6

3.   Typically 5-7 pairs of legs; two pairs of antennae — crabs, shrimp, lobster, copepods (Class Crustacea)

Body oval and flattened; seven pairs of legs, In damp places such as logs, mulch, crawl spaces — pillbugs and sowbugs (Order Isopoda); page 220

Body long and narrow; 10 or more pairs of legs; one pair of antennae 4

4.   Body segments flattened, with one pair of legs per body segment — centipedes (Class Chilopoda); page 202

Body segments cylindrical, most with two pairs of legs per body segment — millipedes (Class Diplopoda); page 218

5.   Abdomen joined to the cephalothorax (head and thorax combined) by a slender waist; segmentation of abdomen indistinct or absent — spiders (Order Araneae); pages 288-295

Abdomen broadly joined to cephalothorax; abdomen distinctly segmented and lengthened to form a long tail with a stinger at the tip — scorpions (Order Scorpionida); page 303

Abdomen broadly joined to cephalothorax; abdomen unsegmented and without a long tail with a stinger at the tip — ticks and mites ..... (Order Acari); pages 136-141, 150-153, 204, 214, 258

6.   Winged 7

Wingless 17

7.   Front wings, at least partly hardened, leathery or parchment-like at base 8

Front wings membranous 13

CHAPTER **4**

PCT Technician's Handbook

8.   With chewing mouthparts 9

     With sucking mouthparts 12

9.   With pincers (cerci) at tip of abdomen — earwigs (Order Dermaptera);.
     page 206

     Without pincers at tip of abdomen 10

10. Front wings hard and shell-like, without veins, may have lines — beetles
     (Order Coleoptera); pages 208, 216, 226-243, 254-257, 260-263, 266-273

     Front membranous wings with branched veins 11

11. Jumping insects, hind femur enlarged; tarsi four or fewer
     segments — crickets (Order Orthoptera); pages 210-213

     Walking insects hind femur not enlarged; head concealed from above by
     pronotum (first dorsal thoracic segment); body flattened and elongate oval
     — cockroaches; pages 166-187 (Order Blattodea)

12. Front wings leathery at base and membranous at tip — true bugs (Order
     Hemiptera); page 200

     Front wings of uniform texture — cicadas, plant and leaf hoppers (Order
     Homoptera)

13. Two wings and two halteres (knob-like structures) — flies and mosquitoes
     (Order Diptera); pages 134, 154-165, 188-199

     Four wings 14

14. Wings typically covered with scales; mouthparts long and coiled under head
     — moths and butterflies (Order Lepidoptera); pages 244-253

Wings with few or no scales; mouthparts not long and coiled 15

15. Hind wings equal to or larger than front wings; wings held flat over abdomen;mouthparts close to eye; all legs walking type; cerci at tip of abdomen short (two to eight segments) — termites (Order Isoptera); pages 274-283

Hind wings smaller than front wings 16

16. Tarsi (last segments of leg) composed of two to three segments; chewing mouthparts — booklice (Order Psocoptera); page 222

Tarsi composed of more than three segments (typically five); some with . chewing mouthparts — ants, bees and wasps (Order Hymenoptera); pages 110-131, 284-287, 296-302

17. Antennae present; collophore (tube-like structure on abdomen) present; spring-like structure usually present — springtails (Order Collembola); page 224
Both collophore and spring-like structure absent   18

18. Two or three long tail-like appendages present — silverfish (Order Thysanura); page 264

Long tail-like appendages absent 19

19. Body flattened laterally or dorsoventrally (like a pancake) 20

Body not flattened; abdomen and thorax narrowly joined together — ants (Order Hymenoptera); pages 110-131

20. Body flattened laterally — fleas (Order Siphonaptera); pages 142-145

Body flattened dorsoventrally (like a pancake) 21

21. Sucking mouthparts externally visible; antennae longer than head — bed bugs and other true bugs (Order Hemiptera); page 132

    No sucking mouthparts externally visible 22

22. Antennae longer than head; first segment of tarus not swollen; tarsus composed of two to three segments; tiny insects — booklice (Order Psoccoptera); page 222

    Antennae shorter than head; head narrower than the thorax at the point of attachment to the thorax — sucking lice (Order Anoplura); pages 146-149

CHAPTER 5
Equipment
Tools of the Trade
Care and Maintenance

The acceptance and increased emphasis by the industry on IPM programs have prompted the development of hundreds of new industry tools as well as changes to many items that have been industry-standards for years. Most of the new products have focused on nonchemical approaches to pest management, typically in the areas of exclusion, mechanical, and biological control.

Pesticide application equipment has undergone major revamping with the development of baits and application devices, sprayers designed for application of small amounts of product deep into harborage, and changes in dusters and aerosol generation equipment. Regardless of why these changes have occurred, the current trend within pest management is to reduce the risk of pesticide exposure.

EXCLUSION STRATEGIES

Products designed to prevent pest entry and eliminate harborage sites are

Caulk is widely used by technicians to seal pests out of a structure.

used more frequently in IPM programs than any other nonchemical products. Many companies have realized that most pest problems within the structural environment originate outdoors and that it is easier and more environmentally responsible to stop them before they enter structures, thus, avoiding the use of pesticides indoors.

Caulk is one of the most widely-used exclusion materials within the industry. It is used indoors in order to eliminate harborage sites and outdoors to eliminate points of entry around pipes, wires, win-

EQUIPMENT

dows, siding, doors and other wall and roof penetrations. There is a vast choice of caulking materials with different formulations (e.g., silicone, latex, etc.), expandability, slow and fast-drying, colors, adhesive qualities, and paintability properties.

Products selected for use must adhere to the application surfaces and have the customer's approval regarding application, color and paintability. Under most circumstances, caulk should be used to fill small gaps and holes that are about 1/2-inch or smaller in size. If larger gaps require repair a filler material, expandable caulk, and other exclusion material should be used.

Screen, wire gauze, sheet metal, hardware cloth, and similar materials are used when more permanent repairs are necessary. The type of product used depends on the size of the pest (e.g., small mesh screening is effective for flying insects but does little to stop rats and mice; likewise, 1/4-inch hardware cloth prevents rodent entry but is ineffective in stopping insects.). The composition of metal products is particularly important outside where these products can rust if exposed to adverse weather conditions such as rain, snow, freezing temperatures, etc.

The most commonly used materials include galvanized metal, aluminum, and stainless steel. Because it does not rust, copper wool should be used instead of steel wool. Screens typically are manufactured from aluminum and nylon which are very durable; however, some insects are able to chew through plastic screens. Cement can be used to seal larger gaps and holes in structures. Regardless of the material used to seal an opening, it should be permanently installed in a manner that forms a tight pest-proof seal.

Seals, door sweeps, and chimney caps are specifically designed to prevent pests from entering structures through fabricated openings. Seals of various composition are used around doors, garage doors, and windows to keep wind and moisture out of the structure. Some products offer the additional protection of preventing the entry of rodent and insect pests.

Door sweeps look like a long brush installed on the bottom of exterior doors. They allow the door to swing freely, yet, when installed correctly, form a tight pest-proof seal. Chimney caps are designed to fit over the top of the chimney in order to exclude birds and other animals such as racoons, and squirrels, which might attempt to use it as a nesting site.

CHAPTER **5**

MECHANICAL CONTROL

This pest management technique utilizes a variety of devices to trap, remove, and/or kill pests. These products use baits, light, and pheromones (i.e., chemicals used by insects to communicate between the species) to attract the pest to the device.

Sticky traps are designed to monitor pest infestations; in some situations, they are used for limited control. Small cockroach infestations inside vending machines, computers, and other sensitive equipment and areas may be handled using sticky traps. Traps may also be used around doorways and entry points in order to capture pests crawling in from outside until permanent repairs can be made. Sticky traps come in a variety of sizes and some contain food attractants and/or pheromones. Sticky traps frequently are used by pest management professionals for flies, stored product pests, occasional invaders, and cockroaches.

Sticky traps come in a variety of models and styles.

Glue boards are used primarily for rodent control.

Glue boards, with much larger surface areas, are designed to capture rats and mice; they also are effective in capturing insects. Some traps incorporate a food attractant in the glue; others leave the option of baiting up to the technician. The major limitations to sticky traps and glueboards are correct placement, (i.e., along lines where the pest is expected to travel), and dust and moisture which can destroy the adhesive properties of the trap. Glue boards must be inspected frequently for trapped rats and mice; trapped animals must be quickly removed and destroyed.

Snap traps are among the oldest items still in use by the industry and have remained relatively unchanged over the years. Traps consist of a wood or plastic base with a spring-loaded lever and trigger mechanism. The small trap is used for mice; the large one is used for rats. Those models with an enlarged trigger mechanism often are used without bait. Typically, small trigger traps are baited with food or nesting material. Although traps should be clean, they do not need to be sterile for continued use. Damaged or heavily-soiled traps should be discarded and replaced. Traps should be placed in rodent runs perpendicular to the route of travel with the trigger positioned in the run, and they should be located out of the reach of children and non-target animals.

Live traps are used to trap mice, birds and larger animals which, under most circumstances, will be released at a distant location following capture. The most notable exception are mice which usually are destroyed.

Repeating mouse traps consist of a wind-up mechanism and sweep bar activated by a trigger mechanism as the mouse enters the trap. Once activated, the sweep bar quickly pushes the mouse into a holding chamber. The holding chamber can accommodate a number of mice. Live traps for other animals include wire cages with spring doors connected to a treadle which closes the door once the animal is inside. These traps are usually baited with food and placed inside animal runs and feeding sites.

Light traps come in all sizes and shapes but share two basic characteristics: they use light to attract flying insects (and jumping insects, in the case of fleas) and either electrocution or sticky boards to kill or entrap the insects, respectively. Light traps vary in the intensity and frequency of the light emitted which leads to many manufacturer claims regarding which is the better light trap.

If installed correctly, i.e., away from entry ways and at the appropriate height for the target pest, light traps are an effective means of eliminating populations of flying insects in-

Light traps are designed to control flying insects.

doors. Light traps are not recommended for use outdoors because they tend to attract undesirable insects, such as mosquitoes, into the area being protected. Light traps must be serviced on a regular basis to remove the accumulation of dead insects and replace glue boards. Failure to do so can lead to secondary pest infestations such as dermestid beetles which feed on dead insects.

Jar traps (also available with bags) are used to capture flies and yellow jackets. To attract insects, the traps are baited with food materials specific to the insect pest. The traps are designed with entry ports to permit the insect to enter at the top of the trap and pass through a "funnel", thus, preventing the insect from flying or crawling out of the trap. Trapped insects either succumb to exhaustion or drown in fluid at the bottom of the jar. Because spoiled food attractants and dead insects cause foul odors, it is important to inspect the traps frequently.

Vacuums are used to physically remove pests, such as cockroaches, spiders, and other occasional invaders, from structures. They vary significantly in

size and shape, and some are outfitted with HEPA filters designed to remove the smallest particulate matter from the air. Consideration must be given to the method of handling the pests that have been vacuumed into the bag or other filtering device. Disposable bags are easily handled by removing the bag, placing tape over the opening, and discarding it in the trash. Pests collected in reusable bags can be killed by spraying a

Vacuums can be used to physically remove pests, like cockroaches, spiders, and other pests, from an account.

small amount of pyrethrin into the vacuum while it is operating or taping the opening and placing the bag in a freezer for several days.

BIOLOGICAL CONTROL

Biological organisms, by-products of biological organisms, and products which affect the growth and development of pests have been used to manage pest populations since the beginning of time. Several of these natural organisms have been isolated and harnessed for use in order to control a variety of agricultural pests. Unfortunately, this technology has limited use in structural pest management.

Predators and parasites are the most commonly recognized biological control organisms. Unfortunately, they do not play a major role in the management of structural pests. Several parasitic wasp species which attack fly pupae and cockroach egg capsules have been identified and cultured. A lack of customer acceptability is the major limitation to their use indoors because hundreds of thousands of parasites are often necessary to effect control, and they are not self sustaining in the absence of their host insect.

Nematodes are unsegmented, soil-inhabiting worms which are parasitic on plants and animals. Several species have been identified and cultured for use against termites, cockroaches, fleas, and mosquitoes. The greatest success with these organisms has been in controlling fleas and mosquitoes.

Bacteria, single celled organisms, are the oldest commercially-marketed biological pesticide. Bacteria must be ingested in order to cause toxicological affects on the insect. Most pathogenic bacteria contain a crystalline spore and toxin. When the spore enters the insect gut, it dissolves, thus, releasing the toxin which then causes the insect's digestive enzymes to eat holes in its gut, thereby releasing the bacteria and gut contents into the blood. These events lead to an infection which eventually kills the insect. Bacteria in structural pest control is limited to mosquitoes, black flies, and fungus gnats. Avermectin, a bacterial derivative, recently has been developed as a very effective bait for cockroaches and ants.

Fungi are multicellular organisms closely related to plants which cause diseases in a variety of insects. They infect the insect through spores which require moisture and humidity in order to germinate and penetrate the cuticle (hard outer skeleton) of the insect. Infected insects die from toxins produced by the fungus. A fungus, *Metarhizium anisopliae,* has been developed for use on cockroaches and termites.

Insect growth regulators (IGRs) are synthetic chemicals similar to insect juvenile hormones which regulate growth and development of insects. Exposure to small concentrations of these products disrupts insect molting and development of the reproductive system. Some products cause death in the molting process; others cause only reproductive failure in the population. These products do not affect insects in the pupal stage and adults; therefore, other con-

Some termite baits use insect growth regulator (IGR) technology to control termites in structures.

trol measures are needed in order to manage these life stages. These products are used to manage cockroach, flea, stored product pest, termite, and mosquito populations.

Chitin synthesis inhibitors are products which affect the formation of the insect cuticle (i.e., hard outer skeleton). Thus, because insects exposed to these products do not develop normally during the molting process, they die.

APPLICATION EQUIPMENT

These pest management tools are designed to apply specific pesticide formulations, e.g., liquids, dusts, and granules. Very few items of equipment are capable of dispersing more than one formulation. Thus, the selection of equipment is based on the formulation, site of application, and size of the area to be treated.

LIQUID DISPERSAL EQUIPMENT

Aerosol dispensers are pressurized cans with a push button nozzle or specialized equipment designed to generate small particles. In aerosol cans, the pesticide is dissolved in a solvent and then mixed with a gas; the pesticide is dispersed as a fine spray when the nozzle is depressed. Aerosols are ready-to-use products which require no

A mechanical aerosol generator.

further dilution. Because many of the solvents used in aerosols are petroleum-based, these products should not be dispersed near an open flame nor in confined spaces where electronic ignition pilot lights are used.

Ultra-low volume (ULV) and ultra-low dosage (ULD) generation equipment is capable of dispensing droplets which are even smaller than those produced by aerosol dispensers. This type of equipment is designed to disperse very small amounts of highly concentrated pesticides. These generators are either gasoline or electrically powered and range in size from hand-held units to vehicular mounted models. This equipment is used to control flying insects indoors and outdoors, exposed crawling insects, and to flush insects out of deep harborages.

Compressed air sprayers have been the standard tool of the pest control industry for decades. However, in recent years, the industry has begun to wean itself from baseboard spraying and this type of pesticide application equipment in response to customer demands for more targeted pesticide applications.

These sprayers typically have a tank capacity of one to three gallons and are used to apply residual insecticides in and around structures and for treating areas outdoors such as

The compressed air sprayer is used both indoors and outdoors.

foundations, around doors and windows, pet resting areas, etc. The ideal operating pressure for this sprayer is 25 to 50 psi; lower pressure causes the liquid to dribble out of the spray tip; and higher pressure causes the product to splash. There are several types of nozzles available for use with compressed air sprayers, and selection is based on the site of application *(See diagram, page 59)*.

The solid stream tip and the plastic extension crack and crevice tip are designed to apply a fine stream of insecticide directly into cracks and crevices for crawling insect control. The flat spray tip produces a flat fan pattern and is used to make spot applications to floors, walls, and areas where pests are present. The hollow cone tip produces the largest volume of insecticide and is used for outdoor applications such as treating bushes, foundations, bodies of water, and other surfaces where higher volumes of product are needed.

Hydraulic sprayers are used to treat large areas.

Hydraulic sprayers are designed to allow technicians to deliver large volumes of pesticides at higher pressures. This equipment can be outfitted with either a boom or single nozzle. Only the latter piece of equipment is used in structural pest control operations. The nozzle sprayer is used to treat large surface areas around and on buildings, trees, ornamentals, turf, and to inject pesticides into the soil for the control of subterranean termites.

DUST AND GRANULAR DISPERSAL EQUIPMENT

Bulb dusters are commonly used indoors to dispense dusts for the control of cockroaches and other crawling insets in areas where liquids can not be used,

A bulb duster being used for an exterior application at a house.

e.g., electrical junction boxes and around equipment. The rubber bulb reservoir (four to eight ounces) capacity is fitted with either a screw cap or stopper, and has either a metal or plastic extension tip. Placing a marble or ball-bearing in the reservoir keeps the dust from caking and helps in dispensing fine puffs of dust. A modification of this duster which consists of a bulb duster at the end of a long pole permits the treatment of wasp nests in high places and from a safe distance.

Plunger dusters and *foot-pump dusters* are designed to apply dusts for the control of ticks, mites, and rodents in outbuildings and outdoor sites. The plunger duster has a smaller capacity than the foot duster which can hold one to five pounds of dust. Both dusters can be outfitted with a rubber hose to disperse dust deep into rodent burrows.

Compressed air dusters are adaptations of compressed air sprayers designed to apply only dusts. The dust is placed in the tank which is pressurized using a hand pump or compressed air source, and the dust is dispensed through a hose and nozzle modified for this formulation. One of the best uses for this type of equipment is to dust large void areas such as crawl spaces and attics.

Granular spreaders are of two designs — the push-type spreader which drops granules from a hopper onto a dispersal wheel which evenly distributes the granules in a six to eight foot swath and the granule spreader which is hand held and operates by turning a crank on the side. The capacity of the hand held unit is significantly less than the push-type which can disperse several hundred pounds of granules per hour. These spreaders are used to disperse pesticide granules outdoors to control pests which occasionally invade structures.

Backpack sprayer-dusters have dual capabilities, i.e., they can be used to disperse liquid and dry formulations. They have the additional utility of being carried on the applicator's back and facilitate treatment of isolated ar-

Backpack sprayers can be used to treat large areas.

eas for mosquitoes and areas around structures where it is often desirable to apply larger quantities of pesticides.

BAITING TOOLS

Containerized baits are one of the most popular items in commercial pest management and the over-the-counter consumer market. Formulated primarily for the control of ants and cockroaches, the baits are contained in child-resistant plastic containers. Some of the bait stations have a see-through lid which permit visual inspections to determine the amount of bait remaining. To work effectively, containerized baits must be placed in the trails used by the target pest population.

Bait guns are designed to use either refillable cartridges or pre-filled bait cartridges, and dispense a variety of active ingredients. The gun is outfitted with a syringe-like needle which is used to place small amounts of bait, usually no larger than the size of an "o," directly into cracks and crevices where cockroaches hide. It is much better to have numerous small placements than one large one, and the bait should never be applied like caulking material in a long bead.

Bait guns are a common IPM tool.

A tamper-resistant rodent bait station.

Tamper-resistant rodent bait stations are designed to prevent children and non-target animals from being exposed to rodenticides. These stations are designed to dispense liquid, block, and/or loose bait formulations. In most circumstances, it is preferable for the technician to secure the bait station so that it can not be picked-up by children, pets or other non-target animals. Technicians should always make sure the station is secured and locked.

CARE AND MAINTENANCE OF EQUIPMENT

Pesticides should not be stored within dispersal equipment for extended periods of time. A good rule-of-thumb: liquids can be stored overnight if they will be used the next day, and dry formulations should be stored for no more than a week. After each use, clean and repair pesticide application equipment as needed. Preventive maintenance will minimize problems on the job site. Liquid application equipment should be washed with soap and water after each use and when using different products.

Gaskets, strainers, and filters should be kept clean and serviced on a regular basis. Strainers and filters should be rinsed with water and cleaned with a soft bristle brush. If this does not remove the residue, the nozzle should be soaked in an appropriate solvent. Nozzles which can not be ad-

Equipment should be regularly cleaned and inspected.

equately cleaned or are damaged should be replaced. Valves should be inspected regularly for serviceability, i.e., they should open easily and rapidly, shut-off quickly and tightly, and be replaced if defective.

Plunger and piston cups should be inspected for wear and tightness of fit. Whenever pressurized equipment is cleaned, the leather cups should be lubricated with a light grease.

Nozzles should be cleaned very carefully to avoid any damage to the opening in the tip. Rinsing with soap and water is usually sufficient, however, heavy residues can be removed with a soft bristle brush and by soaking in an appropriate solvent. If the nozzle opening is worn, it should be replaced.

Hoses should be inspected on a regular basis for cuts, cracks, bulges, and fraying and replaced if a defect is found. Couplings should be inspected for leaks and repaired.

Dusters should be cleaned weekly by removing excess and caked dust from the chamber. The dust tube should be cleaned with a stiff brush and caked dust should be carefully removed from filter screens and small openings.

All electrcially powered equipment should be cleaned and inspected for any defects in the cord and extension cords. The grounding pin should not be removed, and all outlets used during normal operations should be grounded. Electric cords, when worn, should be replaced.

Gasoline powered equipment requires minor maintenance, i.e., the oil level should be checked, and the air filter should be cleaned monthly. Oil changes should be based on the hours of use and the manufacturer's recommendations but no less than annually at the beginning of each season. At the end of the season, the remaining gas from the tank should be drained and the engine run until all the fuel in the carbuerator is gone. Pumps, tanks, and hoses stored outdoors in freezing temperatures should be drained and antifreeze put in the pump. Equipment manuals, which are the sources of information for repair parts and care and maintenance instructions, should be stored in an accessible area.

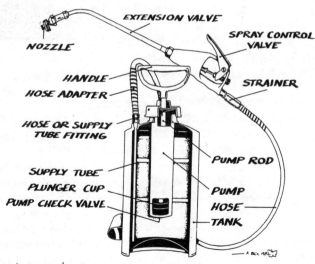

A compressed air sprayer.

CHAPTER 6
Safety

General
Laws and Regulations
Labeling

Pesticides
Application and Handling
Spill Prevention and Cleanup

Webster defines *accident* as "a happening that is not expected" and *safety* as "freedom from danger or injury." Most of the accidents that occur within structural pest management are usually the result of carelessness, failure to take the time to do the job right, and/or failure to consider the consequences of one's actions. The pest management technician's most important responsibilities are customer and personal safety.

Many hazards exist within the pest management industry, e.g., exposure to human diseases, falls, electrical hazards, working in enclosed spaces, etc. Technician's will be exposed to some, if not all, of these hazards as they provide pest management services.

When working within health care facilities, e.g., hospitals, long term care facilities, etc., technicians often are concerned about exposure to infectious diseases such as TB, AIDS, and hepatitis. While working in a patient's room, the risk of exposure to these diseases is thought to be minimal if precautionary measures are followed. Usually information is posted on or outside the door if a patient has a communicable disease. If the pest management technician must enter the room of a patient with a communicable disease, the necessary protective measures should be obtained from the nursing staff.

Within health care environments, the greatest risk of technician exposure is via contact with infectious wastes, particularly blood and excretory-contaminated materials, syringes, and other contaminated medical equipment. These materials should never be handled by the pest management technician. It is especially important not to reach blindly under cabinets and into areas of poor visibility because a simple prick from a carelessly discarded needle could be disastrous.

Patients also are at risk of exposure to diseases carried by technicians. The doors outside patients' rooms and other sensitive areas always should be checked for information regarding protective measures *before* entering the area. Preventive measures might involve no more than the wearing of a surgical mask; they could, however, require donning full gown, head cover, gloves, and shoe covers. It is important to note that some of these same precautionary measures are required within food and drug manufacturing facilities and many medical research laboratories.

Electrical hazards often are taken for granted and not given the respect they deserve. The pest management technician should exercise caution when using three-prong wall outlets and not assume that they are grounded. This is a common problem in older construction but also can occur in newer structures. There are only two ways to assure safe grounding of electrically-operated equipment. The least expensive is a three-prong lighted plug which is inserted in the outlet with a light to indicate if the outlet is grounded. The other is to use a ground fault interrupter which offers protection in the event that the outlet is not grounded or an electrical line is hit.

Liquid pesticides must *never* be used in or around electrical equipment. To prevent severe electrical shock when dusting around electrical outlets and switch plates, a plastic extension on the tip of the bulb duster should be used.

Fogging and aerosol treatments are used occasionally to knock down flying and/or exposed crawling insects. The formulations used in these applications contain 3% or less active ingredient; the remaining solvent material usually is a petroleum distillate. When using these products, air exchangers, fish tank aerators, pilot lights, and electronic ignitions on stoves, hot water heaters, and furnaces *must* be turned off. The consequences of overdosing with total release aerosols while failing to follow these simple precautions have led to disastrous explosions and, in some cases, destruction of the structures. If too much product is dispensed or applied too close to plastics, fabrics, and wall coverings, damage, e.g., staining, etching, etc., can occur.

Occasionally pest management work requires entry into enclosed spaces such as sewers, attics, crawlspaces, silos, and unventilated storage areas. The Occupational Safety and Health Act specifies safety measures which must be taken when entering enclosed spaces. In some cases, two people must be present, and specific safety equipment and various other precautionary measures may be required.

Often entry into areas which contain insulation materials, such as asbestos and fiberglass, is required. Fiberglass, the most common insulation material currently used, is a respiratory and skin irritant. When working in attics and other confined spaces, a dust respirator, coveralls, gloves, goggles, and hat should be worn. This is even more important during the summer as sweating causes more fiberglass fibers to adhere to the skin. Upon exiting the area, exposed skin and protective equipment should be thoroughly washed.

At times, pest management services will require the use of a ladder for climbing onto sloped roofs. If it is necessary to work more than 10 feet off the ground upon a ladder, an assistant should *always* be present to stabilize the ladder and to provide assistance in the event of a fall. The positioning and stabilization of the ladder is important regardless of the height at which one is working. A ladder used to access a roof should extend at least three feet above the top edge of the roof. The base of the ladder should be positioned one fourth the length of the distance from the base of the ladder to the point that it contacts the structure away from the base of the structure.

Foot and back injuries are other safety issues which often are overlooked. When working with heavy items that might be dropped, one's feet should be protected with steel toe shoes or boots. Proper lifting techniques should be used to avoid back strain and injury. When working around construction and renovation sites, steel soled shoes should be worn to protect the feet from nail puncture wounds.

SAFETY AND ENVIRONMENTAL PROTECTION

Service technicians occasionally work with pesticide products that, if used or handled improperly, pose potential risks to the environment, e.g., company work area, the customer's residence or place of business, the natural environment, etc. It is important to remember that the environment is composed of **all** living and nonliving things.

LAWS AND REGULATIONS

Many laws and regulations exist that affect service technicians and their daily activities. Service technicians rarely are expected to read and interpret laws and regulations; they are, however, responsible for complying with them. The Federal Insecticide, Fungicide, and Rodenticide Act (FIFRA) is the most comprehensive law within the United States concerning the use of pesticides. It regulates pesticide registration and labeling, certification and training, enforcement, and other aspects of pesticide handling and use.

FIFRA states that, "no one may sell, distribute or use a pesticide unless it is registered by the EPA." When the Environmental Protection Agency (EPA) registers a pesticide, it also approves the product's label and labeling and classifies the product as general-use or restricted-use. Restricted-use products may be applied only by certified applicators or under their direct supervision.

FIFRA categorizes pesticide applicators as either commercial or private. Each category has specific certification requirements. Private applicators use or supervise the use of restricted-use pesticides on property that is owned or leased by them or their employers for the purpose of producing an agricultural commodity. Commercial applicators include all other certified applicators who apply or supervise the use of restricted-use pesticides.

Each state is responsible for the operation of training and certification programs for private and commercial applicators. In order to become a certified commercial applicator, most states require experience as a pesticide applicator in addition to taking a written exam. Commercial certification is offered in several categories depending on the state. Periodic recertification, a process which often requires attendance at approved training programs and/or retesting, is required by the state.

The most important aspect of the FIFRA, as it relates to service technicians, concerns pesticide application, i.e., the FIFRA requires labeling and is the basis for enforcement. *The label is the law.* It is a violation of the FIFRA and the law to "use any product in a manner inconsistent with its labeling."

The Federal Food, Drug and Cosmetic Act (FFDCA) establishes tolerances or acceptable levels of certain pesticides in food. FIFRA also regulates, through labeling, residues in food. Both Acts are applicable to private and commercial applicators who apply pesticides to or around raw and/or processed food products.

The Food, Agriculture, Conservation, and Trade Act (FACT) requires private and commercial applicators to record the use of restricted-use pesticides. However, the state and/or the pest management technician's company may require records for other or all pesticide use. Records of restricted-use pesticide applications, which must be maintained for two years, must include: product name, amount applied, date applied, location of application, and size of treated area.

The Hazardous Materials Transportation Act (HMTA) concerns transportation of hazardous materials. The Department of Transportation is the federal agency responsible for implementing these regulations. Only a few products used in pest management, i.e., fumigants, and reportable quantities of 2,4-D (100-lbs), diazinon (1-lb), chlorpyrifos (1-lb), and pyrethrins (1-lb), are classified as hazardous materials. Depending on the type of hazardous materials that is being transported, special training and a commercial driver's license may be required, and the vehicle may require marking and/or placarding. If reportable quantities of hazardous materials are transported in a vehicle, the law requires that shipping papers be carried in the vehicle.

The Occupational Safety and Health Act (OSHA) requires employers to provide a reasonably safe workplace and to inform employees of potential health hazards associated with their jobs, e.g., pesticide application. OSHA also requires employers to have a written hazard communication program, a material safety data sheet (MSDS) for every hazardous product used, and to provide training on hazardous materials.

The Endangered Species Act (ESA) regulates activities which can affect endangered or threatened animal and plant species. Because the use of some pesticide products can affect threatened and endangered species, their use may be limited within some geographic locations. Whether use of an active ingredient is prohibited or limited is determined by individual counties.

LABELING

According to the Federal Insecticide, Fungicide, and Rodenticide Act (FIFRA), a pesticide label is the written, printed, or graphic material on, or attached to the pesticide or device or any of its containers or wrappers. Labeling is defined as "all labels and all other written, printed, or graphic matter accompanying the pesticide or device at any time; or to which reference is made on the label or in literature accompanying the pesticide or device."

It is against the law (i.e., the FIFRA) to "use any pesticide in a manner inconsistent with its labeling." Technicians must be able to read and understand the product label in order to apply the product. The law establishes strong penalties (i.e., fines and/or imprisonment) for anyone who misuses a pesticide.

The cardinal rule regarding use of pesticides is the label is the law. The label contains all the essential information relevant to the use of a pesticide product. It provides specific information regarding sites at which the product can be used, which pests will be controlled, mixing instructions, and safety information which the pest management technician is expected to know and understand. Before using any pesticide product, the label must be read and understood. Any questions or doubts should be resolved before the application is made.

Pesticides are classified as restricted-use if they can cause adverse effects on the environment even when applied according to the label directions. Adverse effects include potential injury to the applicator. It is the use of the product that is restricted, not the active ingredient. Restricted-use pesticides can be purchased only by certified applicators and applied by them or under their supervision.

General-use pesticides are products determined by the United States Environmental Protection Agency not to cause adverse effects on the environment. General-use pesticides are considered to be safe for use by uncertified individuals. Most pesticide products are classified as general-use. States have the authority to further restrict the use of pesticides, even when EPA has decided not to do so. A state may classify a product as restricted-use even though EPA has registered it as a general-use product.

Technicians should always read the product label carefully before applying any pesticide product.

All states require that private (e.g., farmers) and commercial (i.e., individuals applying *any* pesticide for commercial purposes not on their own property) applicators must satisfy certain training, certification, and licensing requirements. Consumers do not require any training to apply general-use pesticides.

The following information outlines the important parts of a pesticide label that a pest control technician needs to be familiar with before treating an account. Technicians should always read the label thoroughly before making an application:

Classification. All pesticides are classified as either restricted-use or general-use. A pesticide classified as a restricted-use pesticide must so state. A statement that a product is general-use is not required.

Product Name. This is also referred to as the trade or brand name which is usually registered by the manufacturer with a trademark; it is the name usually seen in advertising, the one most commonly used and recognized in the industry, and the most prominent word(s) on the front panel.

Formulation. This indicates the form of the product in the pesticide container, e.g., D = dust, EC = emulsifiable concentrate, WP = wettable powder, G = granule, ME = microencapsulated, etc.

Ingredients Statement. This area lists in table form the active and inert ingredients and synergists as well as their respective percentage compositions in the product.

Net Contents. The total weight or volume of the material in the container is usually expressed in terms of pounds or ounces for dry formulations (D, WP, G) and in gallons, pints, ounces or liters for liquid formulations (EC, ME, S).

Signal Words. These words provide information on the toxicity of the product to human beings. Every pesticide is required to have a signal word prominently displayed on the front panel, and it always follows the statement, "Keep Out of Reach of Children."

The signal words DANGER-POISON printed in red with a skull and crossbones indicates a highly toxic pesticide. If misused, this type of product is very likely to cause acute injury from inhalation, ingestion, and/or contact with skin or eyes. Products, which are so-classified due only to their potential to cause skin or eye irritation, do not utilize either the word POISON or the skull and crossbones.

The signal word WARNING indicates a moderately toxic pesticide. If misused, this type of product is moderately likely to cause acute injury from inhalation, ingestion, and/or contact with skin or eyes.

The signal word CAUTION indicates a slightly toxic pesticide. If misused, this type of product has only slight potential to cause acute injury from inhalation, ingestion, and/or contact with skin or eyes. Most over-the-counter pesticides fall into the "slightly toxic" classification.

Manufacturer's Name/Address. The name and address of the principal manufacturer must be listed. Often a telephone number is included.

EPA Registration/Establishment Numbers. All pesticide products must be registered with EPA and are, thus, assigned a registration number. The establishment number indicates the facility at which the product was produced. This area of the label usually lists the manufacturer's address.

Precautionary Statements. This section of the label lists special areas of concern regarding the use of the product as well as hazards to humans, domestic animals, and the environment. This part of the label also lists physical and chemical hazards including fire and explosions. Acute effects' statements provide information regarding potential health effects associated with inhalation, ingestion, and/or contact with the product. It also indicates personal protective measures and equipment to be used. The label may provide information about long term health effects and reactions, such as skin irritation or asthma, that are associated with product use. Environmental statements may refer to protection of: ground water from contamination or runoff; bees, fish or birds from exposure; as well statements regarding endangered species.

Directions For Use. The statement, "It is a violation of Federal Law to use this product in a manner inconsistent with its labeling," is always found in this section of the label as a reminder that the label is the law. The directions for use are usually the most lengthy and detailed part of the label. They contain information regarding sites of application, pests controlled, mixing instructions, application directions, and special considerations.

Directions may require knowledge of other definitions, e.g., crack and crevice, spot and broadcast applications, food handling, and nonfood areas of food handling establishments, how to calculate square footage or acreage for outdoor applications, and square footage and cubic footage for indoor applications.

The product label can limit the rate of application and/or prescribe the maximum application pressure. Directions can contain a reentry statement which indicates when a treated area can be safely entered by the customer or technician, e.g., the surfaces are dry, the structure has been ventilated, or after a specified period of time.

Storage and Disposal. All pesticide labels contain information on the proper storage and disposal of pesticides and their empty containers. Label language can best be described as either mandatory or suggestive. Mandatory words include must, do not, avoid, etc., whereas suggestive language includes should, may, recommend, suggest, etc. When in doubt as to the meaning of label language, the pest management technician should check with his/her supervisor or assume that the language is mandatory.

Every pest control company is required by the Occupational Safety and Health Administration's (OSHA) Hazard Communication Standard to establish and maintain a written hazardous communication program (HCP) for their workplace and employees. A major component of this program is material safety data sheets (MSDSs) for each pesticide product used by the pest control company. The MSDSs are usually provided by the manufacturer of the product.

The MSDS is not part of the product labeling. However, often there is essential information on the MSDS which pesticide applicators should know and understand if they will be handling the product. The MSDS discusses the product in terms of the concentrate, i.e., the way the pest management company receives products from the manufacturers and prior to dilution. OSHA does not require MSDSs for diluted products other than those shipped ready-to-use. The MSDS contains the product name and hazard summary, ingredients, physical data, fire and explosion hazard data, reactivity data, a health hazard assessment, spill or leak procedures, special protection information, and regulatory information.

PESTICIDE SAFETY

Pesticides are products which are designed to kill or manage pests by affecting their behavior or growth, e.g., repellents, insect growth regulators, plant growth retardants, etc. They are an important component of integrated pest management. The selection of a particular pesticide depends on the site, pest, active ingredient, and formulation.

Pesticides can be classified in one of three ways:

• According to the type of pest controlled, e.g., insecticides control insects; avicides control birds; piscicides control fish; herbicides control weeds; acaricides control mites; fungicides control fungi; rodenticides control rodents;

etc. This is the most comon method of classification.

   • According to the stage of the pest affected, e.g., when insecticides affect eggs, they are known as ovicides; when they affect larvae, they are known as larvacides; and when they affect adults, they are known as adulticides.

   • According to their mode of action, i.e., how they affect the pest. Stomach poisons are ingested into the digestive system, contact poisons kill through absorption, and respiratory poisons are absorbed when the pest inhales the product.

Unless pesticide products contain 100% toxicant, which is rarely the case, they contain two or more components. The most common are the active ingredients, inert ingredients (i.e., inerts), and

Pesticides should always be stored in a secure, dry area to prevent spoilage.

occasionally a synergist. This combination of components is referred to as a formulation.

   The active ingredient is the component in a pesticide which kills the pest or affects pest behavior. In its purest form, the active ingredient is referred to as a technical grade pesticide. Technical grade pesticides often are too toxic to use as is and so must be diluted.

   Various products, called inerts because they are considered to be non-toxic, are used to dissolve technical grade pesticides, dilute them in water or other carrier, and allow them to be easily used. Some of the inerts commonly found in pesticides are solvents, emulsifiers, spreading and wetting agents, adhesives or stickers, masking agents, carriers and diluents, etc.

   Synergists are products which are used in combination with some active ingredients in order to increase the activity of the pesticide. The combined effect of the active ingredient and the synergist is greater than their sum (i.e., 1+1=3). The most common pesticide formulations are dusts (D), granules (G), wettable powders (WP), solutions (S), emulsifiable concentrates (EC), microencapsulates (ME), and baits.

   Dusts (D) consist of fine particles of dust such as talc or clay which are coated with a thin layer of the finely pulverized active ingredient. Dusts are easy to use, last a long time, and do not stain or injure plants. Most dusts are ineffective when wet, do not stick to vertical surfaces, and are easily removed by rain and wind.

Granules (G) are identical to dusts except that the particles are much larger and usually are formulated using vermiculite. Their properties are similar to dusts, but they do not stick to vegetation, and

Dusts are popular with technicians looking for long-lasting control in dry conditions.

they last longer than dusts. Some granules require water for activation.

Wettable powders (WP) are dry dust formulations designed for dispersal in water. They contain wetting agents (i.e., detergents) and other inerts which help them to mix with water and remain suspended. They do not contain solvents which may have an odor, nor do they affect plants or irritate skin. Wettable powders have the same advantages as dusts, but often they produce visible residues, settle out of suspension, and clog sprayers.

Soluble powders (SP or WSP) are dry formulations which dissolve in water and form true solutions. They are easy to handle and do not require agitation in order to remain suspended. They do not produce visible residues, nor do they clog sprayers or settle to the bottom of the sprayer.

Flowables (F or L) are liquid formulations composed of finely ground particles of the active ingredient. Flowables readily mix with water to form easily-handled suspensions, require frequent agitation in order to remain in suspension, and may leave a residue.

Solutions (S) are liquid formulations which dissolve readily in water or petroleum-based solvents such as kerosene. Solutions do not settle out and do not require agitation.

Emulsifiable concentrates (EC or E) are the most widely used formulations in the pest management industry. They are composed of a liquid active ingredient, petroleum solvent(s), and an additional component, i.e., emulsifier, which is needed in order for the product to mix with water. Emulsifiable concentrates

Solutions dissolve readily in water when mixing.

are easy to handle, remain in solution without agitation, and rarely leave a residue. They tend to be more concentrated, thus, increasing the risk of exposure. Solvents may have an odor and/or affect products such as rubber and plastics which soften when contacted.

Microencapsulated (ME) formulations consist of a liquid or dry active ingredient surrounded by a plastic coating. The formulation is usually applied in water, and the deposited microencapsulated particles gradually breakdown slowly releasing the active ingredient. Microencapsulation allows the active ingredient to remain for a longer period of time than emulsifiable concentrates, maintains the concentration at low levels, and protects it from environmental effects. These products are safer than emulsifiable concentrates for technicians to handle but do require constant agitation in order to remain suspended in the sprayer.

Aerosols are usually ready-to-use products which contain a solvent and a propellant. They dispense very fine pesticide particles into the air for a short period of time. The active ingredient concentration typically is low (e.g., <3.0%). Toxic effects are increased through the addition of synergists.

Ultra low volume (ULV) and ultra low dose (ULD) formulations are similar to aerosols except that their particle size is much smaller. These products are applied using special ULV equipment; they do not utilize a propellant. An ULV (ULD) active ingredient is highly concentrated,

Aerosols dispense very fine pesticide particles into the air for a short time.

often approaching 100%; however, the quantity of material applied is very small, often no more than one ounce per 1,000 cu. ft.

Fumigants are gases which, under the proper conditions, readily penetrate all areas within a space, and, in some cases, the materials that are being fumigated. Fumigants are used

Fumigation treatments penetrate all areas of a structure and can be used against drywood termites and stored product pests.

to treat stored products, e.g., grain and packaged foods, soil, entire structures, vehicles, etc. Fumigants must be contained because they readily dissipate into the atmosphere because they are lighter than air. They are the most toxic substances currently used in pest management, and, thus, require special personal protective measures in order to guard against inhalation.

Baits are formulated as gels, granules, blocks, dusts, and pastes, all of which contain attractants and toxicants. Most baits are food-based; however, other attractants are currently being developed. Baits contain a low amount of active ingredient and remain available to pests for extended periods of time. The drawback to using some baits is that they are attractive to children, pets, and nontarget animals. In addition, some become stale and easily infested with nontarget pests and must compete with alternative food sources. Dead pests often cause odor problems and/or attract other scavenger pests. Technicians who apply pesticide products are responsible for ensuring that individuals, their property, and our environment are protected.

PERSONAL PROTECTIVE MEASURES

The population at greatest risk for pesticide exposure is not customers, their children, or pets but, rather, the pest management service technician who applies the products. Exposure can result during mixing, application, transportation, or by accident. In most cases, however, exposure can be prevented by using safe handling procedures and personal protective equipment (PPE).

The hazard of working with a pesticide product is related to two factors: toxicity and exposure (i.e., HAZARD = TOXICITY x EXPOSURE). Toxicity is determined by the route of exposure and usually is measured using lethal dose (LD = mg/kg body weight) or concentration (LC = ppm or mg/l). Toxicity is reported as the amount of toxicant required to kill 50% of the test population (i.e., usually rats), and it is expressed as LD_{50} or LC_{50}.

The smaller the LD_{50} or LC_{50} the more toxic the product. For example, a product with an LD_{50} of 256 mg/kg is 10X more toxic than one with an LD_{50} of 2,560 mg/kg.

When working with a particular pesticide product, very little can be done to control the toxicity portion of the hazard equation. However, the potential hazard will be influenced by the selection of the product, e.g., Category III products are the least toxic, and their use can reduce the hazard.

The greatest degree of control that can be obtained with regard to the HAZARD equation concerns exposure. This is applicable to customers, children, pets, nontarget organisms, and the individuals at greatest risk of exposure, pesticide applicators. The best defense against pesticide exposure is the use of personal protective measures.

Pesticide applicators usually are exposed to pesticides via contact with skin, ingestion and/or inhalation. Pesticide applicators are more likely to be exposed to pesticides through skin or dermal contact. Most exposure occurs on the hands and arms because hands are used to handle concentrates, dilute products, and make applications.

Ingestion of pesticides usually occurs when pesticide applicators fail to

The proper protective gear should always be worn when applying pesticides.

wash their hands after handling products and then proceed to eat. Inhalation of pesticides usually occurs during application of an aerosol, mist, vapor, etc. and/or by smoking after handling a pesticide product. Inhalation can be avoided by the wearing of a respirator.

Two Federal laws, the Occupational Safety and Health Act and the Federal Insecticide, Fungicide, and Rodenticide Act, establish regulations that are designed to protect pest management technicians from pesticide exposure. Most of the regulations developed from the OSHA involve respiratory protection. Depending on the type of respiratory protection, testing for proper

USEPA classifies products into three categories based on toxicity:			
Category I	highly toxic	$LD_{50} \leq 50mg/kg$	DANGER-POISON
Category II	moderately toxic	$LD_{50} > 50-500mg/kg$	WARNING
Category III	slightly toxic	$LD_{50} > 500-5000mg/kg$	CAUTION

fit and physical examinations may be required. Care of and maintenance requirements for respirators are also included in the regulations.

All labels contain precautionary statements, many of which relate to personal protective measures required when using the product. Precautionary statements on the label are either mandatory, using directive statements and words such as must, or suggestive, using words, such as should. Pesticide applicators are responsible for reading and interpreting all the information on a product label including, but not limited to, precautionary statements.

Examples of precautionary statements on a commonly used structural product include: "May be fatal if swallowed," "Harmful if absorbed through the skin or inhaled," "Causes eye irritation," "Avoid breathing dust or spray mist," "Avoid contact with skin, eyes or clothing," "Wash thoroughly with soap and water after handling and before eating, drinking or using tobacco."

The best way to avoid breathing dust and spray mist is to wear a respirator when mixing and applying the product(s). In order for the pest management technician to avoid pesticide contact with the skin and eyes, coveralls or uniforms, chemical-resistant gloves, and goggles should be worn. Precautionary statements may be specific to the operation being performed or site of application, e.g., during mixing, application in confined spaces, application outdoors, etc. In the absence of definitive statements such as "wear a respirator," company policy and common sense should be the guide.

Personal protective equipment (PPE) includes respirators, goggles, coveralls or uniforms, hats and other headgear, chemical-resistant gloves, and footwear, which are available in various sizes, shapes, and materials.

Contact with the skin is the most likely means of exposure to pesticides. Thus, the best method of protection is

Precautionary statements on the label often outline what safety gear is needed.

to wear gloves, coveralls or uniforms, and protective footwear. Headgear is optional but always should be worn when making overhead aerosol and liquid applications.

PPE should consist of chemical-resistant materials. Cotton, leather, canvas and other woven materials are not chemical-resistant even to dusts which might penetrate the fibers and remain through several washings. Cloth-lined gloves, aprons, boots, and headgear with leather or fabric sweat bands should not be worn; although they are comfortable, once contaminated with pesticides, they are impossible to clean and too expensive to replace.

Products fabricated from rubber or plastic, such as butyl, neoprene, polyvinyl chloride, and non-woven fabric coated with a barrier material are resistant to dry and water-based pesticides. The packaging surrounding protective equipment should be checked to determine if the materials are chemical resistant or chemical or liquid- proof.

Non-water-based product solvents might react with a variety of materials that are resistant to water-based products. The PPE may discolor, blister, soften, dissolve, crack or become stiff. If any of these problems should occur, the item should be discarded and replaced with a chemical-resistant product.

Pesticide labels may require service technicians to apply pesticides while wearing chemical-resistant gloves and, occasionally, footwear. However, the technician's hands and feet can still be contaminated with product unless the protective gloves and footwear are chemical-resistant, worn correctly, in good repair, clean, and replaced periodically.

Most indoor applications of pesticides can be made while wearing sturdy shoes and socks. Canvas shoes are not recommended. Chemical-resistant knee boots should be worn when making outdoor applications to turf, ornamentals and other low vegetation.

Steps must be taken to avoid contamination of the inside of gloves and footwear which can occur when gloves are removed to adjust equipment, open a container, or move or wipe something, and then donned before washing the hands. Contamination of gloves and footwear will be avoided if the technician washes them with soap and water *before* they are removed and then washes and dries his/her hands. Contamination of protective equipment can also occur when pesticide is allowed to run into the gloves and/or boots. This is preventable by wearing overalls, and by creating seals around the gloves.

If the majority of work will be overhead, sleeves should be tucked into the glove. If the work is at or below waist level, the sleeve should be placed

Long-sleeve shirts and gloves are recommended equipment for technicians applying pesticides.

over the glove. Some gloves and coveralls have straps that allow a tight seal to be formed; if not, tape can be used to form a protective seal that will prevent the product from running into the glove. When wearing knee-high boots, the pant legs of the coveralls should be placed over the outside of the boot.

Hoods and wide brimmed hats can be used to keep products from dripping or settling on the head and neck. Plastic bump caps and safari type hats are good choices. Baseball caps are not recommended because they usually are made of fabric and so may absorb product.

Because eyes are very sensitive to pesticides and readily absorb products, they must be protected. Goggles, face shields, or safety glasses with shields over the brow area and the sides should be worn. Safety glasses or tightly-fitting full face respirators should be worn when applying mists, aerosols and fogs. Regular eye glasses do not offer adequate protection from pesticide exposure because they lack shields over the brow area and sides.

A respirator should be worn if the product label states, "do not breathe vapors or spray mist" or if the product is applied within a confined space thereby exposing the technician to vapors or airborne particles.

Labels may specify the type of respiratory protection required as well as when it must be worn. In all cases, the respiratory protective device must be approved by the National Insititute of Occupational Safety and Health (NIOSH) or the Mine Safety and Health Administration (MSHA) for the product being used.

Three designs of air-purifying respirators are available: the dust/mist mask covers the nose and mouth and removes particulate matter; the cartridge respirator consists of one or two cartridges which remove particles; and the canister respirator removes particulate matter.

The most commonly-used respirator within the industry is a half-face cartridge respirator. It is considered to be an air purifying respirator and is effective against pesticide vapors and dusts, but it offers no protection against fumigants and high vapor concentrations. Cartridges and filters should be replaced after eight hours of use, if odors, tastes and/or irritation are detected when using the respirator, or if breathing becomes difficult.

Other types of respiratory protection may be required in pest management operations including self-contained breathing apparatus (SCBA) and supplied air respirators. This protective equipment is required when working with fumigants and in an oxygen-deficient atmosphere.

Respirators help protect technicians from pesticide vapors and dusts.

Respirators are available in various sizes and so must be properly fitted to the individual who will be wearing them. Respirator fit and maintenance are essential if the respirator is to perform properly. Full-face respirators fit more tightly than do half-face respirators. If the seal is not airtight, vapors and dusts will be able to enter the mask, thus, defeating the purpose of the respirator. Individuals with beards will be unable to establish an airtight seal. Respirator fit should be tested by placing the hands over the exhalation valve; when exhaling the mask should inflate slightly until the seal breaks. Fit can also be tested by placing the palm(s) over the cartridge(s) and attempting to inhale; with a good fit, air should not enter the mask.

Before cleaning, the respirator should be inspected for damage and serviceability of flap valves. Prior to washing the respirator the cartridges and prefilters should be removed and stored in a sealed container or discarded if they are no longer serviceable. These items will not perform properly if they are saturated with water. The respirator face-piece should be washed with soap and water,

Respirators must be properly fitted to be effective and safe for technicians.

sanitized with an alcohol swab or soaked in a weak bleach solution for two minutes, dried, reassembled, and properly stored in an air tight container such as a sealable plastic bag to prevent contamination.

At the end of the workday and prior to removal, chemical-resistant items

CHAPTER **6**

should be washed with soap and water. Work clothes should be removed and placed in a separate plastic bag or hamper away from personal clothes and family laundry. Pesticide-contaminated clothing should be washed separately from uncontaminated clothing and laundry or be laundered by a uniform service. Non-chemical-resistant materials, e.g., fabrics, should be laundered by rinsing either in a washing machine or by hand, then washed a few items at a time in hot water with heavy duty detergent. Two complete warm-water rinse cycles should complete the process. The washer should run a full complete wash cycle using hot water and detergent in order to prevent contamination of other laundry. The preferable method for drying fabric products is to allow them to air dry outside for a minimum of 24 hours. A dryer could be used, but pesticide residues may accumulate in the dryer over time.

PPE is either disposable or reusable. When disposable items become contaminated with pesticides, they should be discarded in accordance with the manufacturer's directions. All PPE should be stored in a clean, dry area where it will not be exposed to pesticides or their vapors.

Exposure to pesticides can cause either acute (immediate, i.e., within 24 hours) or chronic (i.e., long term) health effects. The degree to which someone's health may be affected depends on the active ingredient, solvents, formulation, and route of exposure to the product.

Acute effects usually result from pesticide contact with skin, eyes, inhalation, or ingestion. These effects can often be overcome if treated immediately. Symptoms of pesticide exposure are related to the route of exposure; however, many of these same symptoms are characteristic of diseases and exposures to other products in our environment.

Personal protective equipment should be stored in a clean, dry location.

A pesticide that has been swallowed can result in burns to the mouth, throat, and stomach. Inhaling a pesticide can result in severe irritation to the respiratory tract and lungs. Both routes of exposure can result in absorption into the blood and circulation throughout the body.

Pesticide contact with the skin can result in redness, irritation, blistering, and cracking. Eye exposure can result in irritation, temporary or permanent blindness. Skin and eye irritants often cause redness and/or rashes; more severe reactions include blisters and burns. Exposure of the eyes, nose, and throat can result in tearing, redness, swelling, stinging, and burning.

Swallowing, breathing, and excessive absorption of a pesticide may result in sweating, chills, increased thirst, difficult breathing, chest pain, muscle aches and cramps, tremors, and convulsions.

Anyone who experiences acute symptoms suggestive of exposure to a pesticide should seek medical help immediately. Most problems can be resolved if treated early. If the symptoms are life threatening, e.g., breathing has stopped, dizziness, fainting, heart palpitations, etc., life support and first aid procedures, i.e., cardiopulmonary resuscitation (CPR) should be instituted and medical assistance should be sought immediately.

If exposure symptoms occur, medical attention should be sought immediately.

If there is an exposure to pesticides but there are no life threatening symptoms, the individual should be protected from further exposure, the label read for emergency instructions, and general first aid procedures implemented.

First aid for skin exposure is as follows: Clothing should be removed and laundered. Exposed skin should be washed with mild soap and water. Clothes should be changed and the individual should be wrapped in a blanket to avoid becoming chilled. If skin is burned or blistered, it should be covered loosely with a clean cloth or bandage. Ointments, greases, powders, etc. should not be applied to the skin.

If pesticide gets in the eye(s), it (they) should be gently washed for 15 minutes with clean water. Eye wash solutions and other chemicals which might aggravate the injury should not be used. If a pesticide is inhaled, the victim should be moved to fresh air. If the individual is having difficulty breathing, emergency assistance should be called, and, if necessary, CPR initiated.

If a pesticide is ingested, the mouth should be rinsed with water and the individual should drink at least a quart of milk or water. Vomiting should be induced *only* if directed by the label since it may create more problems than it solves if the product is corrosive, an emulsifiable concentrate, or oil-based.

The active ingredient and formulation may play a role in the development of symptoms; however, the symptoms and duration of exposure should determine the need for further medical assistance. Any pesticide incident should always be reported to a supervisor.

APPLICATION AND HANDLING TIPS

Storage, transportation, mixing, and application of pesticides are very important aspects of pest management. The improper performance of these activities can significantly impact upon the environment, health, and property. The pesticide storage site should be secure at all times. Unauthorized persons should be denied access to the storage site which is used only to store pesticides, empty pesticide containers, application equipment, and a spill clean-up kit. Food, feed, seed, fertilizers, gasoline, medical and veterinary supplies, and personal protective equipment should never be stored in the pesticide storage area.

The product label should be checked for special storage requirements, especially those relating to potential fire hazards. Some products are incompatible and, thus, need to be stored separately. All products should be stored in their original containers with legible and easily visible labels. Containers should be sealed. Bait products should be stored in a separate but secure area to prevent contamination with pesticide odors.

The storage area should be checked for leaking containers. If a leak is found, a supervisor should be contacted in order to obtain specific instructions on handling the product. Pesticide products are either transported to the job site diluted and ready for use or in their concentrated form requiring dilution at the job site.

Spills during transport usually occur because of a vehicular accident, broken or leaking containers, and leaking application equipment. If a spill occurs during transport, the leak should be stopped, the spill contained, individuals and sensitive areas protected, and appropriate clean-up procedures instituted.

Pesticides should not be transported in the cab of a truck or passenger area of a van or car. Products transported in these areas may produce odors, spill on individuals, and/or contaminate seats and floor coverings which are virtually impossible to clean. Other products, e.g., food, feed, clothing, that may be contaminated should not be transported.

Pesticides should never be carried in the passenger area of a service vehicle.

Pesticide containers should be tightly sealed and secured in the vehicle so they can not shift or tip over during transportation. Paper containers should be transported so that they do not become wet.

Pesticides should always be secured within the service vehicle even if it will be unattended for only a brief period of time. Pesticides within the trunk, tool box, or other secure location should be locked in order to prevent unauthorized access by individuals, especially children.

Before mixing and applying a pesticide, a product appropriate for the site and problem pest must be selected. The intended site of application must be on the product labeling or it can not be used even if the pest is listed on the label. Even though a site may be listed on the label, other factors should be considered in product selection, e.g., some products may damage surfaces, pose a greater health risk, or damage or kill plants, etc.

Environmental factors which should be considered involve the potential for drift during application, run off, injury to nontarget organisms, ground water contamination, etc. Reportedly, some individuals are sensitive to pesticides and some states require that they be notified prior to application of any pesticide. Technicians must be familiar with the products they are applying and be able to speak intelligently about them with their customers.

Most liquid pesticides that are used in commercial and agricultural pest control require dilution prior to their application. Most dry formulations, e.g., dusts and granules, baits, fumigants, and aerosols are ready-to-use and require no further dilution.

Failure to properly mix a pesticide can result in several problems. Mixing a pesticide that is too strong (i.e., not sufficiently diluted) may kill nontarget animals, leave excessive deposits which may affect human health, stain carpets or walls, cause odors, etc. Some people operate under the misconception that "more is better" so they mix the product stronger than permit-

ted by the label, or they increase the frequency and/or amount of product applied. This is against the law and may result in resistance, repellency, or environmental problems.

It is not a violation of the law to mix or apply a product at less than the labeled concentration or rate unless it is prohibited by the

The proper mixing of pesticides is essential to performing a safe, effective treatment.

label or state regulation. However, this practice is not recommended because the product may not work well and, therefore, will need repeated applications.

Prior to mixing a pesticide, the label must be read in order to determine what type of PPE is needed, how to dilute the product, and how much finished spray is needed for the job. These directions may differ according to the pest, site of application, and the rate of application. Most labels do not provide specific information on how much product to apply to cracks and crevices and spots. Broadcast applications, however, usually specify a certain number of ounces or gallons per 1,000 square feet or acres.

To determine the appropriate amount of a liquid product to apply to a specific site, linear feet and/or square feet will need to be determined. Linear feet can be calculated by measuring the distance around a structure or area. Square footage can be calculated by multiplying the length of the structure or area by its width (A = l x w). If the structure is not rectangular but instead is irregular in shape, it should be divided into smaller squares or rectangles; then the area of each should be caluculated and added together.

Products that are applied as aerosols, ultra low volumes, or fumigants always specify that a certain number of ounces or pounds be applied per 1,000 cubic feet. To determine the volume of a structure, the square footage of one floor should be calculated and multiplied by the height (V = l x w x h). For fumigations of silos calculate the area of the round base (A = pr^2; where p = 3.14) and then multiply by the height.

Calibration of dispersal equipment depends on several factors, the most important of which are the pressure and the size and type of nozzle which will be used for the application. Increasing tank (i.e., pump) pressure and/or nozzle size increases the amount of liquid pesticide discharged at the nozzle opening, and vice versa. Recalibration will be necessary if either of these variables changes.

Many compressed air sprayers are now equipped with pressure gauges so that pressure can be easily monitored and maintained. Resistance to pumping is the only means of maintaining approximately the same pressure when using a compressed air sprayer without a gauge.

After the operating pressure and the nozzle are selected, flow rate can be determined. The sprayer should be pressurized by pumping, and the nozzle opened to collect the discharged product in a measuring cup for one minute. Hint: water should be used to calibrate equipment. The amount of collected material in ounces (this is the flow rate per minute, e.g., 4 oz./minute). Thus, if a label requires application of 8 oz. per 100 square feet, each 100

square foot area requires application for 2 minutes while evenly distributing the spray. Hydraulic sprayers are calibrated in similar fashion; the major differences are pressure and the amount of material applied.

Volumetric applications to enclosed spaces, such as aerosols, ultra-low volume (ULV) applications,

Sprayer calibration is necessary to determine the proper application rate.

and fumigations, also require calibration. An amount of the product should be premeasured into the application unit's tank, the product should be applied for one minute, the remaining amount of material in the tank should be measured, and the difference determined. In most cases, water can not be used to calibrate this type of equipment.

Once the label is read, equipment calibrated, and the amount of finished product for the job determined, it is time to mix, and apply the product. Prior to mixing a pesticide, the precautionary statements on the label should be read in order to determine the appropriate personal protective measures necessary during the mixing. The label should be checked for special mixing instructions, e.g., "fill the sprayer a third full of water, add the appropriate amount of product and add the remaining water."

A mixing site should be chosen, preferably outdoors and away from un-protected people, pets, food, and water sources. An attempt should be made to mix the product on a surface that will not absorb the pesticide if it spills. Using a plastic drop cloth will accomplish this purpose.

Liquid pesticide containers should be placed on a flat surface prior to opening in order to prevent accidental tipping of the container and a resulting spill. Dry formulations in bags should be carefully cut, rather than ripped, open with a knife. The appropriate amount of concentrate should be poured into a measuring device using a tip and pour service container or premeasured dose packs to prepare the finished spray.

All measuring devices and empty liquid pesticide containers should be rinsed using approximately a fourth of the container's volume. The container should be closed and shaken thoroughly in order to coat all sides, and the material poured into the spray tank. This process should be repeated three times. Some pesticide containers can not be rinsed, e.g., bait, dust, granule, and ready-to-use product containers can not be rinsed unless there are specific instructions on the label. Once the pesticide container is rinsed, it must be discarded in accordance with the label instructions. Containers may be recy-clable, refillable, require special handling, crushed or punctured, and put in the sanitary landfill. The label and state disposal regulations should be consulted.

Water sources should always be protected from pesticide contamination by using an antisiphoning device, back flow preventer, or check valve in the water supply line. When filling large pieces of application equipment, a supply pipe with an air gap can be attached to the tank, or a water drop line that has an air gap can be used. These procedures prevent siphoning of pesticide into the supply line.

Mixing pesticides indoors should not occur in an open sink or near a floor drain connected to the sanitary sewer. These sites can be used if the drain can be sealed and if any spilled material can be cleaned up.

Products can only be applied according to label directions. The label may limit application techniques depending on the site, e.g., indoor applications in food processing areas may be limited to crack and crevice and spot, whereas, in nonfood areas, general applications may be permitted.

Prior to application, the label should be rechecked in order to determine what personal protective equipment is needed during application as well as precautions regarding the environment, nontarget animals, unprotected people, etc.

The product should be applied by working away from the area of application. Walking into a treated area should be avoided, but if a treated area must be crossed, appropriate protective equipment, such as rubber boots, should be worn. To avoid drift during outdoor applications, the product should be applied when there is little or no wind.

Some application situations involve a high risk of exposure, e.g., indoor aerosol, fog, and ULV applications, spraying overhead areas, such as ceiling and roof eaves. If such applications increase the risk of exposure, appropriate protective equipment, such as full face respirator, hat, long sleeved shirt, face shield, etc., should be worn.

Once the application is completed, it is important to clean the application equipment because residues may clog and damage the equipment or may be incompatible with other products. Cleaning should be thorough and include the tank, hoses, nozzles, pumps, etc. The equipment should be rinsed with a material compatible with the formulation; in most cases, water is recommended.

The rinse material should be collected, and, if possible, used to dilute mixtures of the same product during future ap-

Technicians should always wash their hands after handling pesticides.

plications. The rinse materials should not be allowed to enter any body of water, drains connected to to sewers or septic tanks, nor to create puddles that can be contacted by children or pets.

After the equipment is cleaned and stored, PPE and clothing should be removed, cleaned as necessary, and hands, face, and other exposed areas of skin washed. it is preferable to shower before leaving the shop to go home. The use of restricted-use pesticides and other products should be recorded as required by state regulation and company policy. In fact, it is good practice for technicians to record the usage information for all pesticide products they apply in an account.

Transportation, storage, mixing, and application of pesticide products is not a job to be taken lightly. The professional technician is responsible for ensurring that these operations are performed safely and in a way that protects people, property, and the environment.

Spill Prevention and Clean-Up

In the normal course of pest management operations, pesticides are occasionally placed into the environment in order to control pests. At other times these products can accidentally be introduced into the environment by spills that occur in storage, during mixing, and transportation. Improper handling of rinse water and disposal of containers may also lead to contamination.

In providing pest management services consideration must be given to sensitive outdoor areas that might easily be affected by pesticides. Water sources, schools, hospitals, sensitvie ornamentals, and the habitats of endangered and threatened species are examples of areas that can be affected by accidents involving pesticides.

Indoor sites are equally, if not more, sensitive than those found outdoors. Sensitive items and areas include children's toys, nurseries, pet bedding and toys, food preparation and storage areas, and indoor plants. In virtually all cases contamination of the environment is an unintentional act resulting from an accidental spill, environmental conditions, improper application, and a lack of awareness of construction and soil conditions.

Environmental contamination can occur in outdoor settings when strong winds cause pesticides to drift off application sites; heavy rains wash the product into surface waters and drains; vehicle accidents and hose breaks cause soil contamination; and fires at pesticide storage facilities cause air and water pollution.

The most serious incidents involving environmental contamination from pesticides result from vehicular accidents. lawn care equipment, termite rigs, and agricultural spray equipment pose the geeatest risk because they transport larger quanitities of concentrates and formulated materials than do most service vehicles.

Indoor environmental contamination often results from a hose break, too much product applied to a surface thereby resulting in run off, too much aerosol dispensed in the air causing settling deposits, etc. In most situations product spills indoors are minor accidents resulting in puddling of materials. Careful observation during application of products will help to minimize the risk of environmental contamination by following application rates, disposal procedures, and other precautionary statements on the product label. Liquid, aerosol, and/or dust products must not be applied near vents that might draw the materials into the air handling system.

Technicians are responsible for the products they use and must be aware of the surrounding environment and should attempt to anticipate what may happen if and when a product is applied. The key to protection of the environment from pesticide spills is prevention. This involves the development of a company spill management plan, training, vehicle and equipment maintenance, and having spill management materials available.

Spill emergency procedures should include indetification of the pesticide, safety and care of the injured, site security, containment and control, reporting, clean-up, decontamination, and disposal. The order in which these activities are performed is determined by company policy and the situation. The company management should, as soon as possible, be informed that a spill has occured. If no one is available, the company emergency response services, such as INFOTRAC or CHEMTREC, should be contacted.

Identification of the products involved in a spill is much easier when shipping papers, product inventory, labels, and material safety data sheets are within the service vehicle. Products should be transported in their origi-nal containers or, if diluted in a service container, they should be marked with the product name, toxicity signal word, name and address of the pest management company and the statement, "Keep Out of the Reach of Children."

During transport, all containers should be tightly sealed and secured in order to prevent spillage.

Product labels and MSDSs should be in the service vehicle for quick reference.

Products should not be transported in the passenger compartment of the vehicle or with other products such as food. Vehicles transporting certain quantities of hazardous materials, such as fumigants, must be marked or placarded.

Every service vehicle should be equipped with an emergency response guide book and a spill control kit appropriate for the type of service provided with the vehicle, e.g., lawn care, termiticide application, fumigation, general household pest control, etc. The emergency response book should contain shipping papers, labels and MSDSs for products being transported, written emergency spill procedures, emergency telephone numbers, first aid and medical emergency procedures.

The spill control kit should include personal protective equipment such as chemical resistant gloves, respirator, eye protection, and coveralls. String, rope or caution tape and stakes to mark off the spill area should also be included. Materials to contain and/or soak up the spill, such as vermiculite, cat litter, sawdust, clay, shredded newspaper, commercial absorbents are used for smaller spills, whereas sand snakes and absorbent pillows or tubes are used for larger spills. A shovel, broom, dust pan and plastic bags or barrels are needed to scoop up the absorbed products. Blank labels should be available to properly label the containers containing absorbed materials. The service vehicle should have a supply of fresh water and heavy duty detergent, a fire extinguisher, and first aid kit.

Each service vehicle should contain an emergency spill control kit.

A handy guide to aid recollection is to use the "three Cs" to manage a pestticide spill, i.e., control, contain, and clean-up. The spill should be controlled at the source of the leak. A compressed air sprayer should be inverted. Vice grips should be used to pinch a broken hose. The pump engine should be turned off. A smaller leaking container should be placed inside a larger container.

Children, pets and other nonessential personnel should be denied access to the spill area. Individuals should remain upwind from the site so that they are not exposed to fumes. The pest management technician should exercise self-protection by donning protective equipment prior to entering the spill area and attempting to contain and clean-up spills. The spill should be contained and prevented from spreading to adjacent areas. Absorbent materials, e.g., commercial absorbents and cat litter, should be used indoors.

Special care must be taken to protect water sources, e.g., wells, ponds, streams, rivers, and lakes. Spilled materials should be prevented from entering sewers, ditches, and floor drains since they ultimately lead to larger water sources. Spills that threaten this critical environment should be blocked or diverted using sand snakes, absorbent pillows, trenching, and/or diking.

Liquid spills should be absorbed with vermiculite, cat litter, sawdust, clay, shredded newspaper, and commercial absorbents. If the spill is on a relatively noabsorbent surface, all the absorbent material should be swept up and placed in a container for disposal. If the contaminated surface is porous, e.g., soil, wood or carpet, it may need removal. Soil that is several inches deep and that covers several square feet may require

Pesticide spills should be contained and absorbed as quickly as possible.

removal. Regardless of the material or quantity, it must be handled and disposed of as an excess pesticide.

Dry spills should be covered in order to prevent the dust from becoming airborne. Dusts should be covered with plastic or sweeping compound and/ or lightly misted with water. Dusts should not be overwetted or they will clump and be unsuitable for reuse. Dry products should be swept and used if still serviceable. Dust products which become too wet or filled with debris should be handled as excess pesticides.

When as much product as possible has been removed, the site should be decontaminated. The label should always be checked for clean-up and decontamination procedures. It is, unfortunate, however, that most labels do not contain this information.

If a spill has occurred on a nonporous surface such as ceramic tile, sealed concrete, sheet vinyl or similar surface, water and a strong detergent solution should be used in order to clean the affected surfaces. The rinse water should be absorbed and the solution prevented from further contaminating the site via running off. Absorbed materials should be swept up and discarded of it as excess pesticides. The spill site should not be left until another knowledgeable person has arrived or until the clean up has been completed.

The service vehicle should be cleaned, as should application equipment and reusable clean-up equipment if they were contaminated during the spill or clean-up process. The vehicle and equipment should be cleaned with chlorine bleach, dish washing detergent, and water. Personal protective equipment should be cleaned per recommendations in the company's policy on care and use of personal protective equipment.

A shower should be taken using plenty of soap and water paying particular attention to hands, arms, feet, face, and any other areas that may have been exposed. Injury and environmental damage can be prevented if hazards are anticipated and proper safety precautions are taken.

CHAPTER **6**

CHAPTER 7
Case Studies
Strategies and Techniques for Effective Solutions

The following are selected "from-the-field" structural pest management case studies and IPM strategies that have proven to be effective solutions. Because each pest management situation is unique, the suggested techniques are not the only means of dealing with the problem. Service technicians are encouraged to consult with their company technical director or supervisor before performing any pest control treatment.

Case Study #1

A thorough inspection of the account is essential.

A large food preparation facility experienced a chronic infestation of German cockroaches. In doing a thorough crack and crevice treatment, the service technician suspected that resistance to the pesticide was the problem. The facility, which was very clean, had four foot ceramic tile walls, a finished concrete floor with several floor drains, and water hoses positioned throughout the food preparation area.

A night time visit to the facility revealed that the majority of cockroaches were exiting harborages around the drop ceiling. Discussions with the facility manager revealed that the kitchen area was washed with high-pressure hoses every day.

An evening service was performed. Accessible cockroaches were removed by vacuuming. Cracks and crevices around the ceiling and other harborage areas were treated with a microencapsulated insecticide, and the facility manager was instructed not to hose down the facility for at least three days. The void area above the drop ceiling was treated with an aerosol pyrethrin formulation (flushing agent). Most of the cracks and crevices were sealed with caulking materials, thereby reducing the number of harborage sites available to cockroaches.

Case Study #2

On a warm winter day, a homeowner complained about a long line of tiny "bugs" trailing down the wall from the corner of the window and that they left a red spot when crushed.

The "bugs" were identified as clover mites. These tiny mites entered the exterior wall void of the house early in the fall to overwinter. The warm winter day was a sufficient cause to stimulate them to come out of hibernation, and they began migrating out of their resting areas.

To avoid stains on the walls, carpets, and fabric items, the clover mites were removed by vacuuming. Exterior wall voids around the window and door frames and wall outlets were dusted with an insecticide and the gaps sealed. It was recommended that the homeowner trim the grass back several inches from the base of the foundation the following spring and that all exterior openings around doors, windows, pipes, wires, siding, etc. be tightly sealed. In late summer, a microencapsulated formulation was applied three feet up the foundation, around doors and windows, and in a band at least ten feet out from the foundation.

Dusts played an important part in this case study's IPM-based program.

Case Study #3

A townhouse owner complained about hundreds of small jumping "bugs" in a downstairs bathroom. There was no previous occurrence of the problem in the five years they lived there.

The "bugs" were identified as springtails. Upon further discussion with the homeowner, it was revealed that the property landscaper had spread mulch around all the town homes just a few days prior to the homeowner discovery of the springtails in the bathroom.

The homeowner was instructed to eliminate all moisture in the bathroom. The mulch was raked back and the soil, and the base of the foundation wall were treated with a microencapsulated insecticide. The mulch was replaced but thinned to no more than a one inch depth. Exterior areas around windows, doors and other wall penetrations were treated with a microencapsulated insecticide.

Case Study #4

Apartment occupants complained about a continuous Norway rat and mouse problem in their building. Traps, bait, and tracking powder previously had been used to kill the rodents. During the last inspection of the apartments there was no evidence of a continuing infestation. Although no rats or mice were seen by the residents, they heard them scratching around in the enclosed ceiling over the kitchen and bathroom.

Further investigation revealed that the area where the sounds were heard housed the vents for the stove and the bathroom exhaust fans. The fans' exhaust ducts were vented to the outside of the building. An outside inspection revealed that sparrows were nesting in almost every exterior vent.

The birds and their nests were removed from the vents, and the nesting area inside the

Bird infestations can be both a nuisance and a potential health risk in an account.

vent was treated with a residual insecticide to kill any bird mites and other pests associated with the nests. The vents were screened with 1/4-inch hardware cloth.

Case Study #5

Four employees working in an centrally-located office within a warehouse suddenly began to experience "bites" on their exposed arms and legs. A thorough inspection of the office and warehouse failed to detect any pest problem. Since the problem was common to all four employees, they were asked about any changes in the office setting, such as new carpet, furniture, copiers, paper products, etc. They reported that they recently had run out of new invoice forms so retrieved some old carbonless paper invoices from the warehouse which they then began to use. The use of the old carbonless forms coincided with the onset of the bites. The carbonless paper forms were replaced with new invoices and the bites disappeared. *Note: carbonless paper and tractor feed forms frequently cause skin irritations in office workers.*

Case Study #6

Homeowners had thousands of minute flies appearing around the windows in the basement of their two-story home. The flies were vacuumed daily, but massive numbers continued to recur.

    The flies were identified as members of the family sphaeroceridae, small flies similar to phorids. Both types of flies breed in decaying organic materials such as raw sewage. A pest control company applied an aerosol insecticide and managed to kill that day's adult population which was immediately replaced the following day. The homeowner attempted to seal the perimeter expansion joint in the basement floor, but very little of the crack was exposed and the sealing was inconsequential. All the drain traps in the house were holding water and there was no detectable sewer odor.

    The homeowners contacted a plumbing company which placed a small camera into the sewer lines under the slab. This revealed that the entire kitchen caste iron waste line had deteriorated, and that raw waste leaked under the slab. Upon breaking up the slab, the raw sewage was writhing with millions of fly maggots. To make matters worse, the sub-slab soil had settled, and the sewage had spread some distance from the kitchen waste line. Pesticide treatment was not necessary. All the contaminated soil was removed, the drain line replaced with PVC, clean fill added, and the floor repaired.

Case Study #7

Approximately 10 years ago in central Florida, a pest control operator found some German cockroaches exhibiting unusual behavior. The population was living outdoors, they could fly, and they were attracted to light.

    Entomologists at the USDA and the University of Florida were contacted,

The application of granular bait was used to control the cockroach infestation facing this technician.

and, upon investigation, discovered that the Asian cockroach, a species identical in appearance to the German cockroach, had been introduced into the United States.

The Asian cockroach is most easily distinguished from the structural dwelling German cockroach by its behavioral characteristics. In contrast to German cockroaches, Asian cockroaches live outdoors; are attracted to light; and fly. The control measures implemented by the technician included exclusion by sealing up exterior entry points around windows and doors, particularly where there was exterior lighting or interior lighting visible at night. Exterior lights using incandescent or mercury vapor bulbs were changed to yellow bug lights and sodium vapor bulbs, and lights were located away from the structure. Populations around structures were baited with a granular scatter bait.

Case Study #8

In late fall, a homeowner found swarming ants in the house. The swarming activity occurred in one corner of a family room addition which was slab on grade construction. The family room was adjacent to the crawlspace under the main house. The homeowner attempted to spray and bait the ants but was unsuccessful. Their point of entry was never located.

Upon inspection by the technician, the ants were identified as large yellow (mois-

Crawlspaces present technicians with a host of potential pest harborage and entry locations.

ture) ants. No activity was found around the exterior of the house. The crawlspace was accessible, somewhat damp, and insulation batts were stapled between the floor joists. There was no apparent evidence of ant activity, so some of the insulation was removed near the corner of the family room where swarmers were found.

This exposed the colony which contained thousands of ants. The ants were in the insulation and the space between the insulation, joists, and subfloor. The ants and nesting site were treated with an aerosol residual insecticide, and the infested insulation was replaced.

Case Study #9

A homeowner had a chronic infestation of Indian meal moths. Several infested food products were located and destroyed, a residual crack and crevice pesticide was applied, and an aerosol application was made. After treatment, the problem continued; after several months, pheromone traps were installed to monitor the situation. A few moths were collected each week until the onset of warm weather at which time there was a significant increase in the trap catch.

During an inspection of the attic nesting material, seeds, crackers and other infestible items were found, apparently brought in through torn screens in the gable vents by nesting sparrows. The attic was monitored using pheromone traps, and a significantly greater number of moths were trapped. The nesting material and contaminated insulation were removed, the gable vent screens were repaired, and a residual insecticide was applied. In addition, the areas around ceiling vents and light fixtures were sealed to prevent moths from entering the living space of the house, particularly around the light fixtures.

Case Study #10

Purchasers of a new home observed holes about the diameter of a pencil in the family room wall as well as an occasional large flying insect with a long tail on the wall. When the insect was finally collected by the homeowners, it was identified as a horntail wasp. It was explained to the homeowners that these wasps occasionally emerge from wooden studs as much as a year or two after construction is completed. The homeowners were reassured that these insects do not reinfest, and thus, no treatment was performed. The holes were patched by the builder and the problem resolved.

Case Study #11

A homeowner, who was deathly afraid of spiders, complained of a chronic problem with web building spiders primarily on the exterior of the home. The initial treatment consisted of dusting the webs with a residual insecticide. Approximately one week following treatment, the remaining spiders and webs were vacuumed. The homeowner was advised that spiders build their webs near exterior lighting and windows illuminated by indoor lights because this is where flying insects attracted to the lights. Exterior lights using incandescent or mercury vapor bulbs were changed to yellow bug lights and sodium vapor bulbs, and lights were located away from the structure.

Case Study #12

Termite swarmers reoccurred for five years, even after repeated treatments with liquid termiticides. The swarmers consistently emerged along the back wall particularly around the rear door frame and a bay window in the kitchen.

An inspection was made in the basement along the band board under the area of activity. While the band board was heavily damaged, there were no live termites. Closer inspection of the siding immediately adjacent to the brick patio which abutted the back of the house for twenty feet in the area of infestation revealed that there was foam insulation running behind the brick and the concrete slab underneath.

Three courses of brick were removed, and the foam insulation behind the brick was exposed. As the insulation was removed, thousand of worker termites were observed to be

Termite work requires knowledge of construction practices.

feeding on the exterior surface of the band board. The insulation extended down behind the concrete slab under the brick. All of the insulation was removed, and the exposed gap and band board were saturated with a termiticide. It was recommended that an asphalt-based expansion material be used prior to replacing the brick and that the brick patio be sealed on a regular basis to prevent water absorption.

CHAPTER 8

Glossary
Terms and Definitions

Abdomen — On insects, the third or last major body region which has spiracles on most segments.

Active ingredient — The component in a pesticide product which kills pests or affects pest behavior.

Adult — A sexually mature and fully grown arthropod which, in most species, is incapable of further growth.

Aerosols — Ready-to-use pesticide formulations which contain an active ingredient, solvent, and propellant.

Allergy — A sensitivity reaction, e.g., asthma, which develops after an initial exposure to a proteinaceous substance.

Anaphylactic — A hypersensitivity reaction to proteins and other substances which, upon a second exposure, can cause life threatening conditions.

Antigen — A proteinaceous substance which causes an allergic reaction.

Antenna (-ae) — The paired segmented sensory structures on the head of an insect; located above the mouthparts and near the eye.

Anterior — In front; before; front.

Arthropod — An invertebrate animal which has a segmented body and jointed appendages; member of the phylum arthropoda.

Bacterium (-a) — Single-celled microorganisms which have no chlorophyll and which multiply by simple division. Some cause diseases and death in insects.

Bait — A pesticide formulation which contains water or food attractant and an active ingredient.

Capitate — A form of insect antenna which is expanded into a head at the tip.

Carton — In Formosan termites, the nest material composed of partially digested wood and soil cemented together with secretions and fecal material.

Caste — In social insects, a group of individuals which have a common functional characteristic, e.g., worker, soldier, and reproductive.

Cephalothorax — Anterior body region in some arthropods in which the head and thorax are fused.

Cercus (cerci) — In some arthropods, a pair of segmented appendages located on the dorsal side and tip of the abdomen.

Chelicera (-ae) — The paired, typically fang-like anterior projections on many arachnids.

Chitin synthesis inhibitor — A chemical which affects the formation of the insect's exoskeleton and causes death during the molting process.

Clavate — A form of insect antenna which is expanded into a club at the tip.

Cocoon — The silken covering over the pupa which is woven by the last instar larva.

Collophore — In springtails, the tube-like structure located on the underside of the first abdominal segment.

Commensal — Refers to rodents which live in close association with humans.

Complete metamorphosis — In insects, the developmental life cycle consisting of egg, several larval stages, pupa, and adult.

Compound eye — This type of eye is composed of many individual cells, each having its own facet on the surface.

Coxa — The first leg segment attached to the ventral surface of the thorax.

Crochets — Small hook-like structures found on the underside of the prolegs on moth and butterfly larvae (caterpillars).

Cuticle — The thin three-layer outer surface of the exoskeleton.

Delusory parasitosis — An individual's imaginary belief that their body is infested with insects.

Deutonymph — In ticks and mites, the second nymphal stage.

Dorsal — Back; upper side; top.

Dorsoventrally flattened — The space between the dorsal and ventral surfaces is narrower than the distance between the two sides; flat like a pancake.

Dusts — A dry pesticide formulation consisting of fine particles of talc or clay which are coated with an active ingredient.

Elbowed — Bent with a sharp angle approximately at 90°.

Elytra — The first pair of wings typical of most beetles which are heavy, shield-like coverings.

Emulsifiable concentrate — A pesticide formulation composed of an active ingredient, solvent, and emulsifier which readily dissolves in water.

Endangered Species Act (ESA) — Regulates activities which can affect endangered and threatened animal and plant species.

Engorge — To fill with blood.

Entomology — The study of insects.

Entomophobia — Fear of insects.

Exoskeleton — The exterior shell or skeleton characteristic of the phylum arthropoda.

Facet — The external covering of a single element of the compound eye.

Federal Food, Drug, and Cosmetic Act (FFDCA) — The law establishing tolerances and acceptable levels of certain pesticides in food.

Federal Insecticide, Fungicide, and Rodenticide Act (FIFRA) — This is the most comprehensive law regulating pesticide use, registration, labeling, applicator training and certification, and enforcement.

Femur — The third leg segment between the trochanter and the tibia.

Festoons — In many hard ticks, the rectangular areas on the posterior edge of the abdomen.

Filiform — A form of insect antenna which typically consists of many segments and is long and thread-like.

Flowable — A liquid pesticide formulation composed of finely ground particles of the active ingredient which mix with water to form a suspension.

Food, Agriculture, Conservation, and Trade Act (FACT) — This law requires private and commercial pesticide applicators to record the use of restricted-use pesticides.

Fontanelle — In some termites, a small pore on the front of the head.

Frass — Solid fecal material; wood fragments typically mixed with excrement produced by wood infesting insects.

Fumigant — A gas which, under proper conditions, readily penetrates all areas within a confined space and kills all life forms present.

Fungus — A plant which does not contain chlorophyll.

Furcula — In springtails, the fork-like structure attached to the tip of the abdomen.

Genal comb — In fleas, a row of very heavy spines projecting downward from the front part of the head.

Granules — A dry pesticide formulation consisting of particles, usually vermiculite, larger than those used for dusts, which are coated with an active ingredient.

Haltere — In flies (Diptera), the knob-like structure which aids in maintaining balance during flight and is attached to the metathorax in place of the second pair of wings.

Hastasetae — Spear-headed setae (hairs) usually found in clusters on dermestid beetle larvae.

Hazardous Materials Transportation Act (HMTA) — This law regulates the transportation of hazardous materials and pertains to very few pesticides.

Head — On insects, the first major body region composed of the eyes, antennae, and mouthparts.

Hypopharynx — In insect mouthparts, the tongue-like structure located in front of the labium and in the center of the other mouthparts.

Hypostome — In ticks, the tube located in the center of the mouthparts.

Inert ingredient — A component of a pesticide which aids in dissolving the active ingredient and facilitates its use.

Insect growth regulator (IGR) — Synthetic chemical analogous to insect juvenile hormones which regulate growth and development in insects.

Instar — The stage between molts.

Integrated pest management (IPM) — A decision-making process that anticipates and prevents pest activity and infestation by combining several strategies to achieve long-term solutions. Components of an IPM program may include proper waste management, structural repair, maintenance, biological and mechanical control techniques, and pesticide application.

Labium — In insect mouthparts, the lower lip located behind the other mouthparts.

Labrum — In insect mouthparts, the upper lip located above the mandibles.

Larva (-ae) — The developmental stages between the egg and pupa in insects with complete metamorphosis. In mites and ticks, the six-legged stage between the egg and nymph.

Larvacide — An insecticide which targets the larval stage of insects with complete metamorphosis.

Mandible — In insect mouthparts, the jaw.

Maxilla (-ae) — In insect mouthparts, the typically jaw-like structure which may have a palp; located behind the mandible.

Mesothorax — The second and middle segment of the insect thorax and the point of attachment for the second pair of legs and the first pair of wings.

Metamorphosis — The development and change an arthropod undergoes from the egg to adult.

Metathorax — The third and last segment of the insect thorax and the point of attachment for the third pair of legs and second pair of wings.

Microencapsulated — A pesticide formulation which consists of a liquid or dry active ingredient surrounded by a plastic coating which gradually releases the active ingredient.

Naiad — The developmental stages between the egg and adult in aquatic insects with simple metamorphosis.

Nematodes — Unsegmented soil-inhabiting worms which are parasitic on plants and animals.

National Pest Management Association (NPMA) — An international/national member-association of pest control operators located at 8100 Oak Street, Dunn Loring, Virginia 22027, 703/573-8330.

Nymph — The developmental stages between the egg and adult in terrestrial insects with simple metamorphosis. In ticks and mites, the eight-legged stage between the larva and adult.

Occupational Safety and Health Act (OSHA) — A law which requires employers to provide a reasonably safe work environment and inform employees of potential health hazards associated with their jobs.

Ocellus (-i) — A small, simple eye composed of a single facet or lens.

Ootheca (-ae) — The purse-shaped egg capsule common in cockroaches.

Ovipositor — In some insects, a long projection which extends from the tip of the abdomen and is used to lay eggs.

Parasite — An organism which lives in or on another organism from which it derives food and shelter.

Pathogenic — The ability of a microorganism to cause disease.

Pedicel — The waist or small segment in ants and some bees and wasps; in ants, it is composed of one or two segments (nodes); in insects, the second antennal segment.

Pheromone — A chemical usually excreted outside the body and used for communication within a species.

Pincers — In some insects, the forceps-like structure at the tip of the abdomen.

Predator — An organism which captures, kills, and eats its prey for food.

Pretarsus (-i) — The sixth and last leg segment located after the tarsus and usually consisting of one or two claws and a pad-like structure.

Proboscis — In some arthropods, the long extension of the mouthparts used for sucking food; a beak.

Proleg — Paired abdominal processes used for locomotion by butterfly and moth larvae (caterpillars) and few other insect larvae.

Pronotal comb — In fleas, the stout spines located along the posterior margin of the first body segment behind the head.

Pronotum — The top or dorsal plate on the prothorax.

Prothorax — The first segment of the insect thorax.

Protonymph — In ticks and mites, the first nymphal stage.

Pupa (-ae) — The developmental stage between the larval and adult stages in insects with complete metamorphosis.

Pyrethrin — A natural insecticide derived from chrysanthemum plants which is irritating to many insect species and is used to flush them out of harborage areas; a toxicant.

Restricted-use pesticide — A product which is determined by the USEPA to cause adverse environmental effects even when used according to the label. These products can be applied only by or under the direct supervision of a certified applicator.

Serrate — A saw-like form of insect antenna.

Simple metamorphosis — In insects, the developmental life cycle consisting of egg, several nymphal or naiad stages, and adult.

Soluble powders — A dry pesticide formulation which dissolves in water to form a true solution.

Solutions — A liquid pesticide formulation which dissolves readily in water or petroleum based solvent.

Spiracle — The exterior opening of the respiratory system.

Stinger — A modified ovipositor which is used to inject venom.

Swarmer — In termites and ants, the winged reproductive stage which leaves the nest in mass to reproduce.

Synergist — A component frequently added to a pesticide product in order to enhance the activity of the active ingredient.

Tarsus (-i) — The fifth leg segment located between the tibia and the pretarsus; often composed of several segments.

Thorax — The middle section of the insect body which is composed of three segments — prothorax, mesothorax, and the metathorax; the point of attachment for the legs and wings.

Tibia — The fourth leg segment located between the femur and the tarsus.

Trochanter — In insect legs, the second segment located between the coxa and femur.

United States Environmental Protection Agency (USEPA) — The regulatory agency responsible for pesticide registration and implementation of training and certification programs for certified pest control operators.

Ultra low dosage (ULD) — In outdoor use, the application of very small quantities of a pesticide, e.g., a few ounces per acre.

Ultra low volume (ULV) — In indoor use, the application of very small quantities of a pesticide, e.g., an ounce per 1,000 cubic feet.

Ultra-Violet (UV) — A wavelength of light which is very attractive to many insect species.

Wettable powder — A dry dust pesticide formulation designed for dispersal in water.

ACROBAT ANTS

Order/Family:
Hymenoptera/Formicidae

Scientific Name:
Crematogaster spp.

Description: Acrobat ant workers are light brown to black and 1/16- to 1/8-inch long. The top of the thorax has one pair of spines, and the petiole has two segments. The petiole is attached to the top of the abdomen in contrast to most other species of ants. When viewed from above, the abdomen appears to be heart-shaped. A stinger is present. When alarmed, the workers scurry around with their abdomens raised over their heads.

Biology: Little has been published on the biology of this species. Swarmers have been observed in nests or swarming from mid-June through late September. The size of colonies range from moderate to large.

Habits: Acrobat ants often nest outdoors under stones, logs, firewood, in trees, and in conditions similar to carpenter ants. In structures, they nest in wall and floor voids, foam insulation, and other areas commonly associated with carpenter ants. They are found in abandoned termite, carpenter ant, or other wood infesting insect galleries. Acrobat ants travel in trails, foraging as much as 100 feet from their nest.

They feed on honeydew produced by aphids and other plant feeders, live and dead insects, including termite swarmers. Workers defend the colony aggressively and are quick to bite and release a foul odor.

Control: Control strategies for acrobat ants are similar to those used for carpenter ants. They require an integrated approach, involving moisture elimination, removal of overhanging tree limbs, stumps, and firewood, and mechanical alterations to prevent entry. All cracks and gaps in exterior walls that provide access to voids or interior areas should be sealed.

It is important to locate the source of the ants, their nests. Areas where water leaks have occurred, particularly roof, soffits, bathroom, and kitchen areas should be inspected. The most complete control is accomplished when the nests are treated with a residual spray or dust. This may involve drilling holes in hollow doors, wall voids, ceiling voids, veneers, etc., which these ants exploit for nesting sites. In some situations, it maybe useful to drill infested wood and apply a dust or liquid formulation directly into the galleries.

Infestations can be reduced but not eliminated by treating travel routes followed by worker ants while foraging. Infestations that originate outdoors can be reduced by the application of barrier treatments using microencapsualted or wettable powder formulations. Baits are of limited value in control of these ants.

Controlling aphids on ornamental plants and trees around structures removes their primary food source and causes them to forage elsewhere for food.

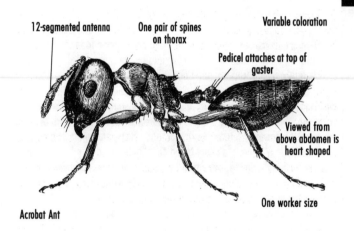

12-segmented antenna

One pair of spines on thorax

Variable coloration

Pedicel attaches at top of gaster

Viewed from above abdomen is heart shaped

One worker size

Acrobat Ant

ARGENTINE ANT

Order/Family:
Hymenoptera/Formicidae

Scientific Name:
Iridomyrmex humilis (Mayr)

Description: Argentine ant workers are 1/16-inch long and light to dark brown, the queens are 1/8- to 1/4-inch long, brown, and covered with fine hair. Males are slightly smaller and a shiny brown-black. These ants lack spines on the thorax which is unevenly rounded when viewed from the side, have a single node on the petiole, and do not have a circle of hairs at the tip of the abdomen. The eyes are located very close to the base of the antennae and appear to be looking forward.

Biology: Colonies consist of several hundred to several thousand workers and several queens. They are located in moist areas near a food source. Developmental time (egg to adult) is 33-141 days and averages 74 days. Swarmers are rarely seen because mating occurs inside the nest.

Habits: Argentine ants are found throughout the southern United States and California. They typically live in nests outdoors near a food source but become major pests when they forage indoors for food. Overwintering nests are large, may have several queens, and are found deep in the soil and in buildings near heat sources. In the spring, these nests disperse with smaller colonies developing in moist soil, in trees, and under stones and concrete slabs.

The colonies are mobile, relocating to more acceptable nesting locations whenever necessary. In the fall, the colonies congregate in communal overwintering sites. Ant numbers decrease somewhat during this period.

Argentine ants are very aggressive and eliminate other ant species in the area they colonize. They attack, destroy, and eat other household pests, such as cockroaches. They prefer sweets, often tending aphids or scale insects on plants, and use them as a source of honey dew.

Control: Outdoors look for Argentine ants trailing up next to the foundation, sidewalks and driveways just below the grass line. All cracks and gaps in exterior walls which provide access to voids and interior areas should be sealed.

Cracks in slabs and gaps in expansion joints should be filled. All debris from the exterior of the structure and other items on the ground where these ants nest should be removed.

Control is often very difficult because the colonies are dispersed. However, the workers are very good at finding sweets in homes and in establishing trails which can be used to find the nests. They are attracted to and feed on sweet and protein-based baits. Baits should be placed where ant trails have been established or in locations where the ants have been sited. Unless using containerized baits indoors, baits should be placed so they are inaccessible to children and pets. Baits should be checked frequently for feeding activity and availability.

The most effective control is accomplished when ant trails are followed to the nests and treated with a residual insecticide. Exterior nests should be drenched with a liquid formulation. Nests in wall voids are more easily treated by aerosol injection or application of a dust formulation. When the nests can not be located, a barrier spray using a microencapsulated or wettable powder formulation should be applied to foundations, plants, and soil immediately adjacent to the building. This is effective in repelling foraging workers and preventing them from reentering the structure.

Controlling aphids on ornamental plants and trees around structures removes a preferred food source and causes them to forage elsewhere for food.

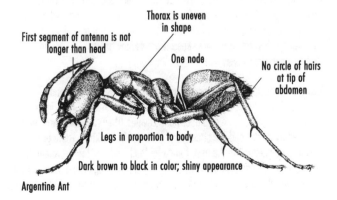

Thorax is uneven in shape

First segment of antenna is not longer than head

One node

No circle of hairs at tip of abdomen

Legs in proportion to body

Dark brown to black in color; shiny appearance

Argentine Ant

CARPENTER ANTS

Order/Family:
Hymenoptera/Formicidae

Scientific Name:
Camponotus spp.

Description: The black carpenter ant, *Camponotus pennsylvanicus,* in the east and *C. modoc* in the west are the most thoroughly studied species in the United States. Other species of *Camponotus* are distributed throughout the country. Carpenter ants are among the largest ants found in the United States, ranging from 1/8- to 1/2-inch long, the queens are slightly bigger. The workers of an established colony vary in size. They are commonly black; however, some species are red and black, solid red, or brown in color. They have one node in the petiole and a circle of tiny hairs on the tip of the abdomen. Their thorax is evenly rounded when seen from the side.

Biology: The adult winged female or queen loses her wings soon after mating with the smaller male and selects a secluded nesting site where she raises the first brood of workers. These workers are very small but assume the care of the larvae and the queen after they mature. Future workers are larger than those from the first brood because they receive better care. All workers are wingless.

Mature colonies range in size from several thousand workers to 10-15,000 including satellite nests for *C. pennsylvanicus; C. modoc* colonies average 10-20,000 workers up to 100,000. When raised at 90⁰ F, black carpenter ants complete their life cycle (egg to adult) in about 60 days. Swarmers do not appear in the colony for several years, usually three to four years for *C. pennsylvanicus* and six to ten years for *C. modoc.* Swarming for these species occurs May through August and February through June, respectively.

Habits: Carpenter ants are social insects that usually nest in wood. They commonly excavate galleries or tunnels in rotting or sound trees and, in structures, readily infest wood, foam insulation, and cavities. They prefer to excavate wood damaged by fungus and are often found in conjunction with moisture problems.

The workers excavate the nest, forage for food, and care for the young. Carpenter ants feed on sugar solutions from honey dew-producing insects such as aphids, sweets, and the juices of insects they capture. They do not eat the wood as they excavate their nests. They actively feed at night well after sunset continuing through the early morning hours. Foraging trails may extend up to 300 feet and, upon close inspection, can be seen on the ground as narrow worn paths.

Carpenter ants enter structures through gaps or cracks while foraging for food. However, the appearance of large numbers of winged adults inside a structure indicates that the nest(s) exists indoors. The workers push wood shavings and pieces of foam insulation out of the nest through slit-like openings in the surface of the wood or other nesting site material. This material, which may contain fragments of other insects, and structural moisture problems are things to look for when trying to locate a colony in an infested structure. Rustling sounds in wall voids are another indication that there is a colony in the area.

Control: Carpenter ant control can be very difficult and thus, requires an integrated approach which involves moisture elimination, removing overhanging tree limbs, stumps, and firewood, and mechanical alterations to prevent entry. It is important to locate the source of the ants, i.e., the nest and satellite nests. Areas where water leaks occur, particularly the roof, soffits, bathroom, and kitchen should be inspected. The most complete control is accomplished when the nests are treated with a residual spray or dust. This may involve drilling holes in hollow doors, wall voids, ceiling voids, etc. which these ants exploit for nesting sites. In some situations, it maybe useful to drill infested wood and apply a dust or liquid formulation directly into the galleries. Infestations can be reduced but not eliminated by treating the trails followed by the workers while foraging. Infestations that originate outdoors can be reduced by the application of barrier treatments using microencapsualted or wettable powder formulations.

Evenly rounded thorax

Variable coloration

Circle of hairs

Usually many worker sizes

Black Carpenter Ant

CRAZY ANT

Order/Family:
Hymenoptera/Formicidae

Scientific Name:
Paratrechina longicornis (Latreille)

Description: Crazy ant workers are light brown to black with a gray sheen and 1/16- to 1/8-inch long. The thorax has no spines and the petiole has one segment. The distinguishing characteristics of this species are their extremely long legs and the first segment of the antenna which is twice as long as the head. The tip of the abdomen has a circle of tiny hairs.

Biology: Little has been published on the biology of this species. The size of colonies tends to be small, containing less than 2,000 workers. A colony of this size may have eight to 40 queens. Occasionally they completely abandon a nesting site and relocate to another.

Habits: Crazy ants often nest outdoors in soil under objects such as trash, refuse, mulch and stones, and in potted plants and cavities in trees and plants. In structures, they nest in wall and floor voids, especially near hot water pipes and heaters.

Their common name relates to their erratic running about in their search for food. They do travel in well established trails, foraging as much as 100 feet from their nest. They feed on honeydew produced by aphids and other plant feeders, seeds, fruit, insects, and almost any household food products. They frequently enter structures in the fall or after a rain because both conditions reduce the availability of honeydew outdoors.

Control: All cracks and gaps in exterior walls which provide access to voids or interior areas should be sealed. All debris from the exterior of the structure and other items on the ground where these ants nest should be removed.

Baits are not the most effective method of eliminating crazy ants because their food preferences change quickly. If baits are attempted, they should be placed where ant trails have been established and in locations where the ants have been sited. Unless using containerized baits indoors, they should be placed so they are inaccessible to children and pets. Sweet baits are the most effective; however, if acceptance is low, consider using a protein bait.

The most effective control is accomplished by following the ant trails, locating the nests, and then treating them with a residual insecticide. Exterior nests should be drenched with a liquid formulation. Nests in wall voids are more easily treated by aerosol injection or application of a dust formulation. When the nests can not be located, a microencapsulated or wettable powder formulation should be applied as a barrier spray to foundations and the soil immediately adjacent to the building. This is effective in repelling foraging workers and preventing them from reentering the structure.

Controlling aphids on ornamental plants and trees around structures removes their primary food source and causes them to forage elsewhere for food.

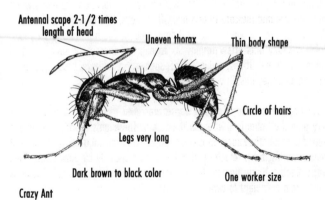

Antennal scape 2-1/2 times length of head

Uneven thorax

Thin body shape

Circle of hairs

Legs very long

Dark brown to black color

One worker size

Crazy Ant

FIRE ANTS

Order/Family:
Hymenoptera/Formicidae

Scientific Name:
Solenopsis spp.

Description: Fire ant workers vary in size, ranging from 1/16- to 1/4-inch long and are yellow to dark red-brown. The thorax lacks spines, and the petiole has two nodes. They have a stinger at the tip of the abdomen and ten-segmented antennae, which is tipped with a two-segmented club.

Biology: The red imported fire ant single-queen colonies range in size from 80,000 - 250,000 workers and 30-100 mounds per acre. Multi-queen colonies have fewer workers per colony but 200-700 mounds per acre. Queens can produce approximately 1,500 eggs per day. The larvae mature into workers in 22-38 days. Minor workers live 30-60 days, intermediate workers 60-90 days, and major workers 90-180 days. Queens live from two to six years. Males die shortly after swarming. Six to eight swarms occur each year and typically contain 4,500 swarmers. Other species have different biological characteristics.

Habits: Several species of *Solenopsis* are called, fire ants, because of the fiery pain their stings inflict upon the victim. These ants usually nest in the ground but can develop colonies in structures, especially in areas near the soil. They are attracted to electrical junction boxes, such as air conditioners and traffic signals. When nesting in the soil, they build large, unsightly mounds which are a detriment to cultivation of fields. Some species of fire ants nest in typical ant habitats, such as under stones, landscape timbers, in voids and around foundations.

These ants, and especially the red imported fire ant, *Solenopsis invicta*, have tremendously large colonies that can severely injure crops, lawns, young birds, and people. Fire ants prefer high protein foods but will feed on practically everything, including other insects, honeydew, seeds, fruit juices, nectar, plants, nuts, cereals, butter, grease, and meats. They also gnaw on electrical wiring and clothing, especially if it is soiled.

Control: Fire ants can cause problems in structures if the workers are habitually foraging in the structure for food. These areas can be partially protected from ants originating in outdoor mounds by applying barrier sprays or dusts of residual insecticides to the soil and foundations around a building. However, this type of treatment is not as effective as treatment of the nest itself.

Nests often require several treatments, especially if they are large and well established. Direct mound injection and drenches can be used successfully to reduce, and in some cases, eliminate the colony. However, baits which contain an insect growth regulator (IGR) and/or a slow acting stomach poison are more successful in eliminating colonies. Seven to ten days after the application of an IGR bait, a residual product should be applied to the immediate area in order to kill foraging workers.

Interior areas should be thoroughly treated with residual insecticides, especially in areas where ants appear to be entering or traveling.

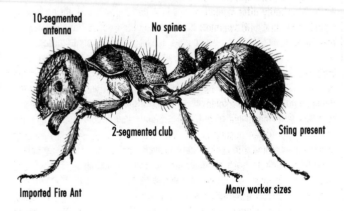

10-segmented antenna — No spines — 2-segmented club — Sting present — Imported Fire Ant — Many worker sizes

LITTLE BLACK ANT

Order/Family:
Hymenoptera/Formicidae

Scientific Name:
Monomorium minimum (Buckley)

Description: The black ant's thorax lacks spines, and the petiole has two nodes. The abdomen is tipped by a small stinger. Each antenna ends in a three segmented club. The little black ant workers are 1/16-inch long. These characteristics also describe the Pharaoh ant; however, the characteristic which distinguishes these two species is the jet black color of the little black ant.

Biology: Little is known about the biology of this species. The colonies have several queens and rapidly grow within a fairly short period of time. The winged reproductives typically swarm in late spring or early fall.

Habits: Little black ants are found throughout the United States. They nest beneath stones, in lawns, and in areas that lack vegetation. Their nests are easily located because they form small craters of fine soil at their entrances. These ants also nest in rotting wood and behind the woodwork or masonry of structures. Indoors they can be found under the edge of carpeting, in old termite galleries, and in wall voids.

Little black ants like to feed on a variety of foods. They eat aphids as a source of honeydew, feed on meats or greases, and are predaceous on other insects. Indoors they feed on both greases and sweet foods.

Control: All cracks and gaps in exterior walls that provide access to voids and interior areas should be sealed. All debris from the exterior of the structure and other items on the ground where these ants can nest should be removed. Firewood should be stored off the ground.

Little black ants are best controlled by locating the nests and treating them with a residual insecticide. Exterior nests should be drenched with a liquid formulation. Nests in wall voids are more easily treated by aerosol injection or application of a dust formulation. Since these ants typically nest outdoors and forage indoors and the colonies are small, it is rarely necessary to apply a barrier spray to foundations and the soil immediately adjacent to the building.

Baits are very effective in controlling colonies. Baits should be placed where ant trails have been established and in locations where the ants have been sited. Unless using containerized baits indoors, they should be placed so that they are inaccessible to children and pets. Sweet baits are the most effective; however, if acceptance is low, a protein bait should be considered.

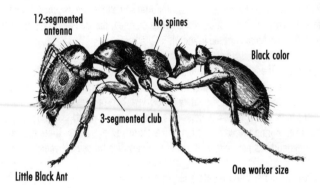

12-segmented antenna

No spines

Black color

3-segmented club

Little Black Ant

One worker size

ODOROUS HOUSE ANT

Order/Family:
Hymenoptera/Formicidae

Scientific Name:
Tapinoma sessile (Say)

Description: Odorous house ant workers are brown to black and 1/16- to 1/8-inch long. The thorax has no spines, and the petiole has one segment which, when viewed from above, is hidden by the rest of the abdomen. There is a slit at the tip of the abdomen instead of a circlet of hairs. The best identifying characteristic is the "rancid butter" smell these ants produce when they are crushed; hence their name. When alarmed, the workers scurry around with their abdomens raised in the air.

Biology: These ants swarm to mate from early May through mid-July, and also mate in the nest, forming new colonies by "budding" off the original colony. A colony has approximately 10,000 workers and several queens, each laying one egg a day. Developmental time (egg to adult) is 34-83 days; however, during the winter, it may take six to seven months. There are several generations per year. Workers and queens live for several years.

Habits: Odorous house ants often nest outdoors under stones, logs, and in the nests of larger ants. They can also nest indoors in wall or floor voids, around heat sources, (e.g., hot water pipes and heaters, crevices around sinks and cabinets). Odorous house ants travel in trails and prefer sweets, although they eat almost any household food. They usually invade structures during rainy periods after honeydew on plants has washed off.

Control: Odorous house ants are very difficult to control. All cracks and gaps in exterior walls that provide access to voids or interior areas should be sealed. All debris from the exterior of the structure and other items on the ground where these ants nest should be removed.

Odorous house ants prefer sweets but are not easily controlled using sweetened baits. Baits should be placed where ant trails have been established or in locations where the ants have been sited. Unless using containerized baits indoors, baits should be placed so they are inaccessible to children and pets.

Sweet baits are the most effective, but if acceptance is low, a protein bait should be considered.

The most effective control is accomplished by following ant trails, locating the nest, and treating them with a residual insecticide. Exterior nests should be drenched with a liquid formulation. Nests in wall voids are more easily treated by aerosol injection or application of a dust formulation. When the nests can not be located, a barrier spray should be applied to foundations and the soil immediately adjacent to the building. This is effective in repelling foraging workers and preventing them from reentering the structure.

Controlling aphids on ornamental plants and trees around structures removes their primary food source and causes them to forage elsewhere for food.

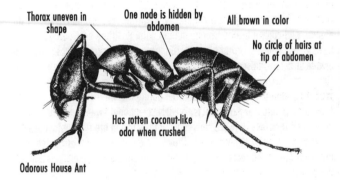

Thorax uneven in shape

One node is hidden by abdomen

All brown in color

No circle of hairs at tip of abdomen

Has rotten coconut-like odor when crushed

Odorous House Ant

PAVEMENT ANT

Order/Family:
Hymenoptera/Formicidae

Scientific Name:
Tetramorium caespitum (Linnaeus)

Description: Pavement ants are 1/16- to 1/8-inch long with a dark body and lighter colored legs. They have two small spines on the back portion of the thorax, two nodes in their petioles, and their bodies are covered with stiff hairs. Pavement ants are easily identified by the narrow, parallel grooves on their heads and thoraxes.

Biology: Little is known about the biology of this species. The developmental time (egg to adult) is 36 to 63 days. Indoors, swarmers emerge anytime, and they emerge outdoors in June and July.

Habits: Pavement ants are commonly found in metropolitan areas in the eastern and central United States and in California. They nest outdoors under flat stones, under sidewalks, along curbing, under concrete slabs, etc. They invade structures in search of food and are a particular problem in areas where slab-on-grade construction is prevalent. Inside structures, they nest in walls, insulation, floors, and near heat sources during the winter.

Pavement ants feed on insects, meats, seeds, and sweets, but they prefer meats and greases. They are slow-moving insects and are frequently observed in areas where they are prevalent. They forage in trails as far as 30 feet from the nest. Although they are not particularly aggressive, workers can bite and sting.

Control: All cracks and gaps in exterior walls which provide access to voids or interior areas should be sealed. Cracks in slabs and gaps in expansion joints should be filled. All debris from the exterior of the structure and other items on the ground where these ants nest should be removed.

Pavement ants are attracted to and feed on sweet and protein-based baits. Baits should be placed where ant trails have been established and in locations where the ants have been sited. Unless using containerized baits indoors, baits should be placed so they are inaccessible to children and pets. The baits should be checked often for feeding activity and availability.

The most effective control is to follow ant trails to the nests and treat them with a residual insecticide. Exterior nests should be drenched with a liquid formulation. Particular attention should be paid to cracks and expansion joints in slabs, driveways, sidewalks, etc. Metal extension tubes should be used to inject liquid products directly into these sites. If subslab areas require treatment, the use of foam injection or liquid subslab treatments should be considered. Nests in wall voids are more easily treated by aerosol injection or application of a dust formulation. When the nests can not be located, a barrier spray using a microencapsulated or wettable powder formulation should be applied to foundations and the soil immediately adjacent to the building. This is effective in repelling foraging workers and preventing them from reentering the structure.

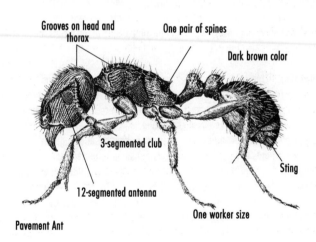

Grooves on head and thorax

One pair of spines

Dark brown color

3-segmented club

Sting

12-segmented antenna

One worker size

Pavement Ant

PHARAOH ANT

Order/Family:
Hymenoptera/Formicidae

Scientific Name:
Monomorium pharaonis (Linnaeus)

Description: Pharaoh ants are very small; workers are about 1/16-inch long. They range from yellow to light brown. The thorax lacks spines, and the petiole has two nodes. These ants can be distinguished from thief ants because they have a three-segmented club at the end of the antenna.

Biology: These ants do not swarm; females mate in the nest, and new colonies are formed by "budding." This means part of the main colony moves *en masse* to a new location. There may be hundreds of thousands of ants in a colony. A female produces 350 to 400 eggs in her lifetime. The entire life cycle is completed in 38 to 45 days at room temperature. Indoors these ants develop year round. Workers live approximately nine to ten weeks, and queens live four to twelve months.

Habits: Pharaoh ants are widely distributed throughout the United States. They can nest outdoors and are at times a crop pest; they are major problems in homes and institutions, such as hospitals, hotels, prisons or apartment complexes. They nest in warm, hard-to-reach locations in walls, subfloor areas, wall sockets, attics, cracks, crevices, behind baseboards, and furniture.

Pharaoh ants eat dead or live insects but seem to prefer meats or greases. They also feed on sugar syrup, fruit juices, jellies, and cakes. These ants are an especially important pest in hospitals where they have been found infesting the dressings on patients' wounds, feeding on secretions from new born infants, in IV tubes, etc.

Control: Because they nest in such a wide variety of locations within structures, Pharaoh ants are very difficult to control. Since they do not necessarily follow specific trails to food sources, they are difficult to trace to their nests; however, it is useful to look near sources of water and food.

A successful Pharaoh ant control program will involve intensive baiting using baits that contain insect growth regulators or slow acting and non-repellent toxicants. Baits should be placed in areas where ants are or are expected to be active. The more placements, the better. Baits must be placed in areas inaccessible to children and pets. Baits should be placed behind outlets, switch plates, areas where wires pass through walls, in cracks and crevices, etc. The baits should be placed along straight lines, such as the edge of baseboards, moldings, etc. In southern states where these ants are active outdoors, exterior baiting programs have eradicated infestations that extended into the structure. The bait should be checked for continued acceptance. If the ants are avoiding the bait, another bait formulation should be used. Baits should be replaced as necessary. Several follow-up visits may be needed each month until control is effected. Elimination of Pharaoh ants from a structure might require a year or more of vigilant baiting.

Residual applications of insecticides should not be used to control Pharaoh ants. These applications affect <5% of the workers in a colony, stress the colony, and cause the colony to split, leading to colonization of new areas with very little reduction in the existing population. It is important that this is explained to customers who might feel compelled to take things into their own hands.

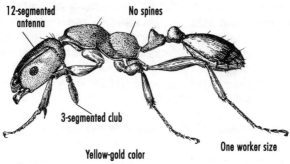

12-segmented antenna

No spines

3-segmented club

Yellow-gold color

One worker size

Pharaoh Ant

THIEF ANT

Order/Family:
Hymenoptera/Formicidae

Scientific Name:
Solenopsis molesta (Say)

Description: Thief ants are very tiny ants; workers are never more than 1/16-inch long. The thorax lacks distinct spines, the petiole has two nodes, and there is a small stinger at the tip of the abdomen. Thief ants are yellow to light brown and look much like Pharaoh ants. These two ants are easily distinguished because thief ants have a large two-segmented club at the tip of the antenna; Pharaoh ants have a three-segmented club. Thief ants also have very small eyes.

Biology: These ants begin swarming as winged reproductives in June; this activity continues until late fall. A colony of a few hundred to several thousand workers can be established by a single fertilized female. Developmental time (egg to adult) is 50 days to several months.

Habits: Thief ants are often found in very large nests that have tiny tunnels connecting to the nests of larger ants. They habitually steal food and brood from the other ants' nests; thus, their name. Thief ants usually nest outdoors in areas with bare soil or under stones. When they do nest in structures, they usually are found in wall voids and similar protected locations. Thief ants feed on live and dead insects, seeds, and honeydew. They will tend aphids and other honeydew-producing insects as a source of this food. They generally prefer food with high protein content.

Control: All cracks and gaps in exterior walls which provide access to voids or interior areas should be sealed. All debris from the exterior of the structure and other items on the ground where these ants nest should be removed.

Thief ants prefer high protein foods; however, they might also respond to sweet baits. Baiting might be ineffective because these ants usually stop feeding on the bait before enough is consumed to eliminate the colony. If baits are used, they should be placed where ant trails have been established and in locations where the ants have been sited. Unless using containerized baits indoors, baits should be placed so they are inaccessible to children and pets. Protein baits are the most effective; however, if acceptance is low, a sweet bait should be considered.

The most effective control is accomplished when ant trails are followed to find and treat the nest with a residual insecticide. Exterior nests should be drenched with a liquid formulation. Nests within wall voids are more easily treated by aerosol injection or application of a dust formulation. When nests can not be located, a barrier spray should be applied to foundations and the soil immediately adjacent to the building. This is effective in repelling foraging workers and preventing them from reentering the structure.

Controlling aphids on ornamental plants and trees around structures removes their primary food source and causes them to forage elsewhere for food.

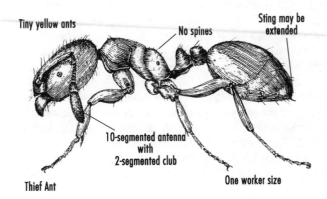

Tiny yellow ants

No spines

Sting may be extended

10-segmented antenna
with
2-segmented club

Thief Ant

One worker size

LARGE YELLOW ANTS

Order/Family:
Hymenoptera/Formicidae

Scientific Name:
Acanthomyops spp.

Description: The most common species, the larger yellow ant, is sometimes called "citronella ant". These ants have a distinct lemony smell when they are crushed. In the Northwest, these ants nest in structural areas where there is high moisture content; hence the common name, moisture ant. They are large ants ranging from 1/4- to 3/16-inch long. They are pale yellow to yellow-red and have a single node in their petioles and a circle of hairs at the tip of the abdomen.

Biology: Very little is known about the biology of these ants. The large winged reproductives develop in the fall and overwinter, emerging in swarms, often by the thousands, in the early spring through early fall. They often emerge into structures (particularly heated basements) causing the occupants to misidentify them as termites because of their size and their appearance during termite swarming season.

Habits: Large yellow ants nest in rotting wood, in the soil, and in the foundations of homes. Indoors they are found in the crawlspace soil, between insulation and subflooring, in moist wood, etc. Outdoors large nests are found under rotting firewood, patio stones, rocks, landscape timbers, etc. These ants tend to excavate large galleries and stack up large amounts of soil adjacent to the nesting site. In some parts of the country, multiple small openings (mounds) may appear throughout the lawn.

They feed exclusively on honeydew obtained from the aphids they tend on plants. Because yellow ants forage at night, they are seldom seen in structures by customers, and perhaps this explains why they have never been reported feeding on human food.

Control: If they are nesting in moist wood in the structure, the moisture source should be eliminated and the wood dried. Rotting firewood that is serving as a nesting site should be removed. These ants are seldom of concern in structures except when the swarmers emerge. Swarmers are best removed using a vacuum cleaner. After collection, the vacuum bag should be sealed and discarded. If this is not practical, a nonresidual aerosol should be used to knockdown the swarmers. Nests located in or around structures should be drenched with a liquid residual product. Baits are ineffective in controlling these ants.

Controlling aphids on ornamental plants and trees around structures removes their primary food source and causes them to forage elsewhere for food.

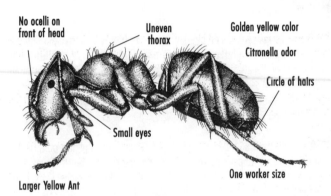

No ocelli on front of head

Uneven thorax

Golden yellow color

Citronella odor

Circle of hairs

Small eyes

One worker size

Larger Yellow Ant

BED BUG

Order/Family:
Heteroptera/Cimicidae

Scientific Name:
Cimex lectularius Linnaeus

Description: The adult bed bug is 3/16-inch long, oval, flat, and rusty-red or mahogany in color. The bed bug is flat and thin when unfed but becomes more elongate, plump, and red when it is full of blood. Four-segmented antennae are attached to the head between the prominent compound eyes. The three-segmented beak, or proboscis, is located beneath the head and passes back between the front legs. The bed bug cannot fly as its wings are reduced to short wing pads.

Biology: As the female bed bug lays her eggs (i.e., one to five per day and 200-500 within her lifetime); she uses a clear substance to attach them in cracks and on rough surfaces. Under ideal conditions, eggs hatch in about seven days and the nymphs molt five times, taking a blood meal between each molt. Development time from egg to adult is 21 days. The adult can live for almost one year.

Habits: The bed bug hides in cracks and crevices during the day, preferring to rest on wood and paper surfaces instead of stone and plaster. It leaves these harborage areas at night to feed on its host which include humans, birds, hogs, and family pets. The blood meal requires three to ten minutes and usually goes unnoticed by the victim. After feeding, the bite site may become inflamed and itch severely in sensitive people. Although the bed bug has been associated with over 25 diseases, transmission has not been conclusively proven. Over time, the harborage areas become filled with the molted skins, feces, and old egg shells of the resident bed bugs. These areas have a characteristic "stink bug" smell caused by a secretion emitted by the bed bug.

Control: A thorough inspection is necessary to detect harborage sites. The odor and specks (i.e., the little spots of excreted blood) that they produce assist in pinpointing these areas. Sanitation is helpful in control of bed bugs because it assists the homeowner to become aware of some of the harborage areas. A vacuum cleaner will remove some of the bugs. Bed bugs can be controlled with thorough applications of residual insecticides applied to cracks and crevices, behind baseboards, and into other known or suspected harborage areas. Furniture, especially mattresses and box springs, should also be lightly sprayed. However, bedding of infants and infirm individuals should not be treated, but, rather, replaced. Dusts can effectively be used in wall voids and attics.

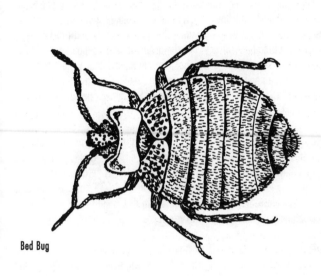

Bed Bug

STABLE FLY

Order/Family:
Diptera/Muscidae

Scientific Name:
Stomoxys calcitrans (Linnaeus)

Description: The stable fly is a 1/4- to 3/8-inch long, light gray fly which is similar in appearance to the house fly. The stable fly has four dark stripes on the back of the thorax (i.e., segments behind the head with the legs and wings attached), but unlike the house fly, it has a light colored spot between the two center stripes. It has a wider abdomen than the house fly and checker-board dark spots on the upper surface. The characteristic which separates the stable fly from the house fly is the bayonet-like mouthparts that project from the underside of the head. The mature larva is a spindle-shaped maggot which is about 3/8- to 1/2-inch long. It has slit-like, "S"-shaped respiratory openings at the end of its body.

Biology: The female stable fly lays her eggs in rotting, fermenting, organic matter including wet hay and straw, accumulated grass clippings, seaweed, manure/straw mixtures, and cannery wastes. The eggs hatch in one to three days, and the larvae complete development in 11 to 30 days. Larvae molt three times before they pupate and remain in this stage for six to 20 days. Under optimum conditions, the life cycle takes 21 to 25 days.

Habits: The stable fly is seldom a pest in houses; however, it is becoming more of a problem because new homes are being built in former agricultural areas. It can be very annoying in outdoor areas around stables, livestock operations, and on some beaches. The adult stable fly uses its piercing-sucking mouthparts to feed on the blood of livestock, pets, and humans. It usually feeds on the lower part of the human body around the ankles. Its bite is very painful.

Control: Control of the stable fly involves sanitation, mechanical control, and the selective use of pesticides. Sanitation includes the complete removal and proper disposal of breeding materials essential for larval development. Waste must be removed at least every two weeks to break the life cycle. Stall muckings should be spread in a thin layer over a field or disposed of at an approved landfill. Accumulations of vegetation, such as grass clippings on golf courses, sea weed on beaches, etc., should be removed.

Screens should be properly maintained and installed to prevent stable flies from entering structures; light traps are effective in eliminating those that do enter. Yellow glue sticks are useful in areas where they are protected from dust and blowing debris.

Adult flies can be controlled by applying wettable powder and microencapsulated formulations to their resting sites. Aerosols, mists, and ULV applications can be used to kill exposed adults.

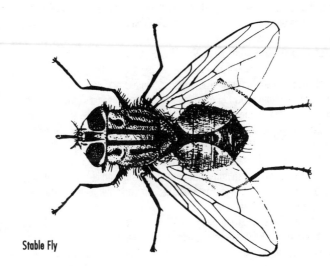

Stable Fly

AMERICAN DOG TICK

Order/Family:
Acari/Ixodidae

Scientific Name:
Dermacentor variabilis (Say)

Description: The adult American dog tick is a hard tick, 1/8- to 3/16-inch long, and red-brown with white markings on the back. The mouthparts are easily seen when viewed from above. The body is flattened and shaped like a tear-drop. The female's shield-like area remains unchanged, but the rest of her body stretches and changes from red-brown to blue-gray as she engorges with blood while feeding. It has 11 festoons (i.e., rectangular areas divided by grooves) along the posterior end of the abdomen. Unlike the adult, which has eight legs, the larva is 1/32-inch long and has six legs and red markings near the eyes, and is pale yellow when unfed. It turns slate gray and doubles in size when engorged. It molts into an eight-legged nymph which is yellow-brown with red markings near the eyes and 1/16-inch long when unfed. It doubles in size when engorged with blood.

Biology: In June or July, the engorged female tick drops off of the host animal to lay from 4,000 to 6,500 yellow-brown eggs in a sheltered location. The eggs hatch in 36 to 57 days. The larva seeks rodents and other small animals for its blood meal. After molting to the nymphal stage, the tick once again seeks hosts. The engorged nymphs molt to the adult stage which usually feeds on dogs and other large mammals. Development (i.e., egg to egg) can be completed in three months, but each stage is remarkably resistant to starvation and the life cycle can be prolonged for up to two years. Unfed adults can live for two to three years.

Habits: This tick is a three host tick, i.e., it requires different and successively larger host animals in order to complete development. It is a very common pest of dogs east of the Rocky Mountains and readily feeds on a variety of other animals, including humans. The American dog tick transmits Rocky Mountain Spotted Fever and can cause tick-induced paralysis if it attaches near the base of the neck.

Larval and nymphal activity begins in March and continues until mid-July.

Nymphs are more abundant during the summer period. Adults usually are active in the spring when they are found in "waiting positions" on vegetation along paths and trails. They attach to passing animals, begin to feed, and mate.

Control: The American dog tick does not survive indoors for long and is seldom a problem in structures except when it occasionally falls off an infested dog. It cannot complete its life cycle indoors. If ticks are found indoors, most of them can be removed by vacuuming and the remainder controlled by applying a residual spray or dust in the areas frequented by the dog. Ticks on the pet should be treated on the same day.

Outdoor control of American dog ticks often is necessary. Debris and ground cover around the area should be removed to discourage rodent and other small animal activity. Rodent populations should be eliminated by using traps, tracking powders, and/or baits. Tick harborage and questing areas (i.e., where the tick waits for a host animal) should be reduced by cutting high grass and weeds. Barrier sprays and perimeter applications of residual pesticides to all vegetation along paths and trails and grassy areas may be needed in order to reduce the magnitude of the problem.

Individuals who work or spend time in tick-infested areas should protect themselves by wearing light colored long sleeve shirts and long pants, tucked into their socks and by treating their clothing and skin with repellents. Attached ticks should be carefully removed so that the head is not broken off and left imbedded in the skin.

American Dog Tick

BROWN DOG TICK

Order/Family:
Acari/Ixodidae

Scientific Name:
Rhipicephalus sanguineus (Latreille)

Description: The brown dog tick is uniformly red-brown and 1/8-inch long when not engorged. Its mouthparts are easily seen when viewed from above. The body is flattened and shaped like a tear drop. The female's shield-like area remains unchanged, but the rest of her body stretches and changes from red-brown to blue-gray and her body size increases to 1/2-inch long, as she engorges with blood while feeding. The male has tiny pits scattered over the back. The adults have festoons (rectangular areas divided by grooves) along the posterior end of the abdomen.

Biology: The engorged female drops off the host dog seeking a quiet location to deposit a mass of 1,000 to 3,000 eggs. She tends to move upward and often lays her eggs in cracks and crevices near the ceiling or roof of kennels and, indoors, around wall hangings and the ceiling. The female dies soon after laying her eggs. In 19 to 60 days, the eggs hatch into six-legged larvae which then move down the wall and attach to a dog as soon as possible. After engorging, the larvae drop off and molt into the eight-legged nymphs, which in turn, molt into the adults after another blood meal. The adults also feed on dogs. The entire life cycle can be completed in two months, and there are usually two to four generations per year. Adults can live unfed for 18 months.

Habits: The brown dog tick is a southern pest but can establish in houses and kennels in more northern areas. The tick almost exclusively is a parasite of dogs but is annoying and frightening to homeowners because often it is seen on walls and furnishings if the dog is infested. It seldom feeds on humans. It is a potential vector of Rocky Mountain spotted fever.

Control: Successful control of the brown dog tick requires a three step program which involves sanitation, tick control indoors and outside, and tick control on the dog(s). The infested house and/or kennel should be thoroughly cleaned in order to eliminate as many ticks as possible. Vacuuming is very effective indoors. Pet bedding and resting areas should be given special attention.

Kennels, dog houses, and structures occupied by pets should be thoroughly treated to control ticks that have dropped off the dog and that reside in harborage areas. Residual sprays and dusts should be applied carefully to all potential tick harborage areas. It is important to remember that these ticks like to reside in the upper portions of structures in cracks and crevices and areas frequented by dogs.

Outdoors, a microencapsualted or wettable powder formulation should be applied to grassy and bushy areas near the house or kennel, the edges of lawns and gardens, under porches, and other areas where the dog travels or spends time.

Infested dogs should be treated by a veterinarian, a pet groomer, or the owner on the same day on which the premises are treated. Pest control technicians should never apply any product to an animal.

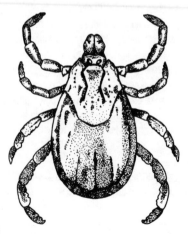

Brown Dog Tick

BLACKLEGGED TICKS

Order/Family:
Acari/Ixodidae

Scientific Name:
Ixodes scapularis Say Deer/Bear
Ixodes pacificus Cooley and Kohls Pacific/Western

Description: Blacklegged ticks are hard ticks, 1/16- to 1/8-inch long, and orange-brown except for the head, shield behind the head, and legs which are dark reddish-brown. Their mouthparts are easily seen when viewed from above. Their bodies are flattened and shaped like a tear drop. The female's shield-like area remains unchanged, but the rest of her body stretches and becomes darker as she engorges with blood while feeding. Blacklegged ticks do not have festoons (rectangular areas divided by grooves) along the posterior end of their abdomens. Unlike the adults which have eight legs, the larvae have six legs and are 1/32-inch long. The eight-legged nymphs are 1/16-inch long when unfed.

Biology: During the winter adults feed on deer. In the spring, engorged females drop off of the host animal and lay 3,000 eggs in a protected area. The eggs hatch in 48 to 135 days, and from June through September the larvae seek and feed on small rodents, such as mice, voles, and chipmunks. After molting to the nymphal stage, the ticks once again seek hosts and feed only once for three to nine days on larger animals such as racoons, opossums, and squirrels. Nymphs are found from April through August of the following season. After 25 to 56 days, engorged nymphs molt to the adult stage which usually feed on deer. Development (i.e., from egg to egg) is normally two years, but in the absence of suitable hosts, development can extend up to four years. Adults live long enough to mate and for the female to lay eggs and then die.

Habits: The deer/bear blacklegged tick is found east of the Mississippi River. The western/Pacific blacklegged tick is found along the west coast and into Arizona. These ticks are three-host ticks, i.e., they require different and successively larger host animals in order to complete development. The larvae and adults commonly infest white-footed deer mice and deer, respectively. However, the nymphs have a much wider range of hosts, including humans. It is this stage which is responsible for the transmission of Lyme disease, the most significant tick-borne disease in the United States. Annually more than 10,000 people are infected with this disease, mostly in the northeast.

Blacklegged ticks climb grass and shrubs and wait for host animals. Typically they are concentrated in transition areas between fields and lower grassy vegetation, along animal trails, and in host animal nests and dens, such as woodpiles, burrows in the ground, stumps, logs, old rat and bird nests, and crawlspaces.

Control: Control of blacklegged ticks involves five steps: sanitation, personal protective measures, inspection, tick control, and rodent control. These ticks are rarely, if ever, encountered, nor do they survive, indoors. Control of blacklegged ticks begins with sanitation. Rodent and other small animal activity should be controlled by removing debris and ground cover around the area. Garbage, pet foods, and other materials which attract larger animals to the area should be cleaned up and securely stored. Tick harborage and questing areas (places where ticks wait for a suitable host to pass) should be reduced by mowing the grass and by removing and cutting high grass and weeds along paths, the fringe of turf areas, and areas frequented by humans and their pets.

Perimeter areas should be inspected for ticks and evidence of rodent activity. Rodent populations should be eliminated by using traps, tracking powders, and/or baits. In perimeter areas, residual pesticides should be applied to rodent runs, nesting areas, vegetation along paths and trails, and grassy areas including lawns, using liquids with spreader stickers, microencapsulated, and wettable powder formulations.

Individuals who work and spend time in tick-infested areas should protect themselves by wearing light colored long sleeve shirts and long pants tucked into their socks. Clothing and skin can be treated with repellents. Individuals should inspect themselves for ticks after returning from tick-infested areas and remove ticks before they attach. Any attached ticks should be carefully removed so that the head is not broken off and left imbedded in the skin.

Blacklegged Tick

CAT/DOG FLEAS

Order/Family:
Siphonaptera/Pulicidae

Scientific Name:
Ctenocephalides felis (Bouche) — Cat Flea
Ctenocephalides canis (Curtis) — Dog Flea

Description: Cat and dog fleas can be found in the same area. They are very similar in appearance. They are small, 1/8-inch long, wingless, laterally flattened, and have piercing-sucking mouthparts. The flea has very well-developed legs allowing it to jump at least six inches straight up. They are black-to-reddish brown. Its body is covered with backward projecting spines which help it to move between the hairs on the host animal. The head of the female cat flea is twice as long as it is high; the head of the female dog flea is less than twice as long as it is high. Both cat and dog fleas have a row of very heavy spines on the front of the head (i.e., the genal comb) and the back part of the first body segment (i.e., the pronotal comb).Cat and dog flea larvae are 1/4-inch long when fully developed. They look much like fly maggots except for their well-developed heads. They have 13 body segments and are dirty-white in color with backward projecting hairs on each body segment. They have a pair of hook-like appendages on the last abdominal segment.

Biology: Cat and dog fleas undergo complete metamorphosis. After each blood meal, females lay four to eight eggs at a time (but 400 to 800 total within her lifetime) on the host animal and/or in its bedding. The eggs fall into the nest and/or bedding of the host animal or wherever the animal happens to be at that time. The eggs hatch in about 10 days, and the developing larvae feed on the adult flea feces which contain bits of dried blood. Depending on temperature, they molt three times in from seven days to several months. When mature, they spin silken cocoons in which they pupate. The pupal stage lasts up to 20 weeks. The adult cat flea often stays within the cocoon until vibrations stimulate it to emerge. Development (egg to adult) requires from 16 days to a year or more.

Habits: Adult fleas feed on blood with their piercing-sucking mouthparts. They typically seek a blood meal within two days of becoming an adult. Cat and dog fleas prefer these two animals but readily feed on other animals, e.g., racoons, opossums, rats, and humans. Adult fleas remain on the host animal throughout their lifetime but are occasionally knocked off the animal by scratching. Occasionally, they can be found in the pet bedding and resting areas. Wild animals nesting in structures can initiate indoor flea populations. Larvae typically are found in areas where pets spend most of their time as well as in animal nesting areas.

Control: Effective flea control requires customer cooperation and involves three major steps: sanitation, insecticide application, and on-animal flea control. The house should be thoroughly vacuumed to remove larvae, pupae, and food materials. The vacuum cleaner bag should be sealed and discarded immediately after vacuuming. Vacuuming lifts the carpet fibers removing debris which, in turn, allows the pesticide to penetrate to the base of the carpet where the flea larvae are found. Pet bedding should be discarded or washed in hot, soapy water.

Indoors, residual insecticides in combination with an insect growth regulator which blocks development of the flea larvae should be applied to carpeted areas, furniture where pets reside, and cracks and crevices on hard floors. Humans and pets should remain out of the treated area until all surfaces have dried and the area has been ventilated. Outdoor areas which are frequented by the pet should be treated at the same time that the house is treated using microencapsulated or wettable powder formulations. These products should be applied either as a spot treatment to areas frequented by the animal or as a broadcast treatment.

The pet should be treated by a veterinarian, pet groomer, or the owner on the same day on which the house is treated. Numerous products are available for on-animal flea control, e.g., pills containing an insect growth regulator, spot-on adulticides, flea collars, on-animal insect growth regulators, soaps, dips, etc. Regardless of the treatment, adult fleas must be eliminated from the animal in order for treatment to be effective. Technicians should never apply any product to an animal.

Cat Flea

ORIENTAL RAT FLEA

Order/Family:
Siphonaptera/Pulicidae

Scientific Name:
Xenopsylla cheopis (Rothchild)

Description: The Oriental rat flea is small, 1/16-inch long, wingless, laterally flattened, and has piercing-sucking mouthparts. It has very well-developed legs allowing it to jump at least six inches straight up. It is black to reddish brown. The body is covered with backward projecting spines that help the flea to move between the hairs on the host animal. The Oriental rat flea differs from cat and dog fleas by not having the heavy row of spines on the front of its head and on the back of its first body segment.

The flea larva is 1/4-inch long when fully developed and looks much like a fly maggot except for the well-developed head. It has 13 body segments, is dirty-white with backward projecting hairs on each body segment, and has a pair of hook-like appendages on the last abdominal segment.

Biology: Oriental rat fleas undergo complete metamorphosis. The female lays from four to eight eggs after each blood meal. Eggs hatch in about 12 days. The larvae feed on a variety of organic debris in the nest of the host, including dried fecal material. They molt three times and reach the pupal stage in 12 to 84 days. The pupal period is spent in a silken cocoon spun by the larvae and lasts seven to 182 days. Fed adults may live for two to four weeks.

Habits: The Oriental rat flea usually is found in association with Norway rat infestations. However, it is found on other animals including other rat species, cottontail rabbits, and ground squirrels. It prefers warm temperatures and is more common in southern states. It transmits bubonic plague and murine typhus to animals and humans. It can jump nine inches horizontally and four inches vertically.

Control: Control of Oriental rat fleas requires a coordinated program which includes sanitation, rodent control, insecticide application to premises, and treatment of infested pets. Sanitation removes food sources for larval fleas, as well as the rat hosts, and thorough vacuum cleaning removes large numbers of the larvae and pupae. The vacuum cleaner bag should be sealed and disposed of immediately after vacuuming. Since rats are the preferred host for this flea, it is important to couple flea control with rodent control. Rodent burrows and routes of travel should be treated with residual insecticide dusts to control fleas on the rats and to prevent them from spreading to humans.

Indoor areas that are infested with Oriental rat fleas should be treated with a residual insecticide in combination with an insect growth regulator to block development of the flea larvae. These products should be applied to carpeted areas, furniture where pets reside, cracks and crevices on hard floors, rat runs and nesting sites. Humans and pets should remain out of the treated area until all surfaces have dried and the area has been ventilated. The outdoor areas which are infested with rodents and which are frequented by pets should be treated at the same time that the structure is treated using dusts in burrows and microencapsulated or wettable powder formulations in other locations. These products should be applied as either a spot treatment or as a broadcast treatment to the entire yard, particularly if stray animals are in the area.

Pets and livestock should be treated by a veterinarian, pet groomer, or the owner on the same day on which the structure is treated. Numerous products are available for on-animal flea control, e.g., pills containing an insect growth regulator, spot-on adulticides, flea collars, on-animal insect growth regulators, soaps, dips, etc. Regardless of the treatment, adult fleas must be eliminated from the animal in order for treatment to be effective. Pest control technicians should never apply any product to an animal.

Oriental Rat Flea

CRAB LICE

Order/Family:
Anoplura/Pediculidae

Scientific Name:
Pthirus pubis (Linnaeus)

Description: As the name implies, the crab louse is crab-like in appearance. It is smaller and broader than the head or body louse, about 1/16-inch long. It has well-formed legs with claw-like tips. The front legs are noticeably more slender than the hind two pairs. The abdomen is broad with lobe-like projections emerging from the sides. The crab louse is a dirty gray color and difficult to see.

Biology: Females lay two to three eggs per day and 15 to 50 eggs in a lifetime, gluing them very securely to individual hairs. The adults and nymphs feed on blood, using their piercing-sucking mouthparts, and move very little once they have settled in place. The nymphs molt three times, and development (egg to egg) requires five to six weeks. Adults live 15 to 25 days but can live only 24 hours off their host.

Habits: Crab lice most commonly infest human pubic and anal regions. Sometimes they will infest the body hair, eyebrows, and facial hair of heavily-

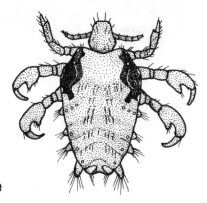

Crab Louse

infested individuals. Crab lice use their pincer-like legs to grasp hairs. Crab lice usually are transmitted during sexual intercourse but they may also be transferred in locker rooms by items such as towels, bedding, and toilets.

Control: Individuals infested with crab lice should contact their physician or public health nurse in order to obtain information on products available for treatment. Since lice do not survive off the human host for any length of time, it is unnecessary to treat the buildings inhabited with lice-infested individuals. Clothing, bed linen, and blankets should be laundered or dry cleaned in order to kill lice that may be on these items.

HEAD/BODY LICE

Order/Family:
Anoplura/Pediculidae

Scientific Name:
Pediculus humanus capitatus DeGeer — Head Louse
Pediculus humanus humanus Linnaeus — Body Louse

Description: Head lice are about 3/16-inch long and wingless, with a distinct head, short, stout antennae, and easily recognized eyes. They have stout, well-formed legs with thumb-like extensions at their tips that help the insects hold onto human hair. Adult lice vary in color from gray-black to dirty-white. Body lice are virtually identical in appearance to head lice except that they are somewhat larger.

Biology: The adult female head louse produces 50 to 100 eggs which hatch in from five to 10 days. The eggs are cemented to hairs very close to the scalp. The immatures have piercing-sucking mouthparts. similar to the adults which they use to feed on the blood of the host. They molt three times, becoming adults within three weeks after eggs are laid. The adults live for 22 to 23 days.

Body lice differ from head lice in that they prefer to feed on the human body instead of the head. The adult female produces 50 to 100 eggs within her lifetime. The eggs are usually glued along the seams of clothing such as underwear and hatch in about five days when the clothing is worn constantly. The nymphal stages are completed in eight to nine days but may take twice that long if the clothing is removed nightly. Body and head lice can survive approximately 48 hours off infested persons.

Habits: Head lice are considered to be the same species as body lice, but they prefer to infest the hair on the head of humans. When removed from this area to other parts of the body, they migrate back to the head.

Body lice are most common in areas where clothing contacts the body, such as the neck, shoulder, armpit, waist, and crotch. They prefer to feed in areas where the skin is soft or folded, such as at joints. They feed using their piercing/sucking mouthparts and remain attached to the clothing even while feeding. Both adults and nymphs are blood feeders.

Head lice are major problems in schools. Children easily transfer head lice to family members and to one another through play and sharing cubbies, hats, coats, combs, etc. Infestations affect all social-economic classes and in no way are an indication of poor personal hygiene or a filthy environment in the home.

Control: Individuals who are infested with head or body lice should contact their physician or public health nurse to obtain information on products available for treating their hair, body and/or clothing. Since head lice do not survive off the human body for any length of time, it is unnecessary to treat the structures which are inhabited with lice-infested humans.

In cold weather, body lice can be killed by removing infested clothing for several days, and subjecting it to freezing temperatures overnight and by laundering it in hot water. Since body lice do not survive off the human body for any length of time, it is unnecessary to treat structures inhabited with lice-infested humans as long as bed clothing, sheets, and blankets are laundered.

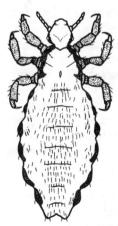

Head Louse

BIRD MITES

Order/Family:
Acari/Various

Scientific Name:
Dermanyssus gallinae — Chicken Mite
Ornithonyssus sylviarum — Northern Fowl Mite
Ornithonyssus bursa — Tropical Fowl Mite

Identification: Bird mites are approximately 1/32-inch long. Adults have eight legs (larvae have six) and the same body shape as ticks. Mites are more transparent, however, than ticks and do not have festoons (rectangular areas on the posterior edge of the abdomen of many hard ticks). The chicken mite has a very broad dorsal shield which gradually tapers toward the posterior end. The northern fowl mite has a very wide dorsal shield for 3/4 of its length and then becomes very narrow. The dorsal shield on the tropical fowl mite is shaped like a tear-drop.

Biology: Chicken mite females deposit eggs in batches of seven or less in crevices and under bird nests. Development (i.e., egg to adult) requires seven days. Fed adults survive up to four to five months without feeding. The northern fowl mite and the tropical fowl mite have similar biologies. The females lay approximately three eggs on the host bird within two days of having become an adult. The development time (egg to adult) is approximately seven days. The northern fowl mite can survive only three weeks off a host bird. Tropical fowl mites survive less than ten days without the host bird and, therefore, pose a short-lived problem in human habitats

Habits: These bird mites readily bite humans. The chicken mite is the most common mite on sparrows, starlings, and pigeons. Of the bird mites, they most often cause human dermatitis. They are intermittent nocturnal feeders and do not remain on the host bird. During the day, they are found in cracks, crevices, and/or the nest.

Northern fowl mites readily bite humans. Though they are usually merely an annoyance, they can, however, cause dermatitis. This mite is a common parasite of sparrows, starlings and pigeons spending most of their time on the birds. They overwinter in bird nests and readily migrate out of the nest when it is vacated.

Habits of the tropical fowl mite are very similar to the northern fowl mite.

They typically are associated with sparrows but prefer to remain in the bird nests instead of on the birds. They do bite humans and can cause dermatitis.

Control: Mites are the most difficult group of ectoparasites to control because they are so small can get through the smallest of openings. When birds leave their nests or the nests are removed, these mites move into other areas of the structure searching for food and water. Because these actions often lead to development of a secondary pest problem, the implications of bird and nest removal should always be anticipated and included as a critical part of the pest management plan. Vacuuming removes many of the exposed mites.

Most potential problems can be solved by removing and immediately disposing of the birds, their accumulated droppings, and their nests. To control bird mites a pesticide should be applied to the nest area and to cracks and crevices in the vicinity of the nest. To kill exposed mites which have spread throughout a room, an aerosol or ULV pesticide should be applied.

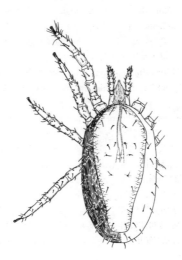

Bird Mite

RODENT MITES

Order/Family:
Acari/Various

Scientific Name:
Ornithonyssus bacoti — Tropical Rat Mite
Laelaps echidnina — Spiny Rat Mite
Liponyssoides sanguineus — House Mouse Mite

Identification: The rodent mites listed above readily bite humans. They are approximately 1/32-inch long. Adults and nymphs have eight legs (larvae have six) and the same body shape as ticks. Mites are more transparent than ticks, their mouthparts are different, and they do not have festoons (i.e., posterior indentations on the rear of many hard ticks). The tropical rat mite has scissor-like chelicerae, narrow tapering dorsal-ventral genital plates, and an egg-shaped anal plate. The spiny rat mite has a very large genitoventral plate with the posterior margin curved inward to accommodate the insertion of the anal plate. The female house mouse mite has two dorsal shields.

Biology: Tropical rat mites complete their life cycle (i.e., from egg to egg) in as little as two weeks. They live approximately 63 days and can remain unfed for five to twelve days. Spiny rat mite females can live up to six weeks. The house mouse mite life cycle (i.e., from egg to adult) requires 18-23 days. They can survive up to two months without feeding.

Habits: Tropical rat mite females and protonymphs (first nymphal stage) suck blood and can inflict very painful bites that result in intense itching and skin irritation. They bite humans even when there are rats present, but biting intensifies when the rats are eliminated. They are not known to transmit any human diseases.

The tropical rat mite is probably the most common mite to infest rodents within the United States. Norway rats are the preferred host for this species. Spiny rat mites feed at night and hide during the day in cracks and crevices or in the host nest. They are found in areas rodents use to nest which are recognized by grease marks and droppings. Spiny rat mites do not transmit diseases to humans. House mouse mite males, females, protonymphs, and deutonymphs suck blood and often cause a localized rash. While the preferred hosts of this species are mice, they also attack rats. They prefer warm harborage areas, such as, pipes, incinerators, and furnaces.

Control: Since almost all rodent infestations can potentially result in a problem with house mouse mites, it is best to institute the following procedures in conjunction with the rodent control program. Otherwise, because the mites are able to survive a week or more without a blood meal, they may show up long after the rodent problem is solved. This may result in a callback.

In situations in which these mites are suspected, a thorough survey should be conducted. Mites crawling across surfaces or on humans can be collected on clear plastic sticky tape and sandwiched with another piece of tape. Sticky traps can be used; however this makes identification very difficult. Mites can be collected by vacuuming with a filter (i.e., a handkerchief or cloth material inserted into the vacuum tube). After removing the filter material, it should be transferred to a sealable plastic bag or alcohol so the mites can be identified.

Rats and/or mice should be eliminated, preferably by trapping. If baits are used, carcasses should be removed. Nests should be removed and the area thoroughly vacuumed. An aerosol provides quick knockdown of exposed mites. Residual products should be applied to nesting areas, runways and other areas where mites are observed.

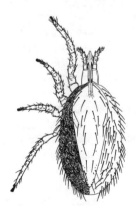

Rodent Mite

AEDES MOSQUITOES

Order/Family:
Diptera/Culicidae

Scientific Name:
Aedes spp.

Description: Mosquitoes are members of the order Diptera, i.e., flies. Mosquitoes are slender, long-legged, two-winged insects that are 1/8- to 1/4-inch long. They are unlike other flies in that their wings and bodies are covered with scales, and they have long piercing and sucking mouthparts.

The larvae, or wrigglers, are aquatic and have well-developed heads, swollen and unsegmented thoraxes, and eight-segmented abdomens which end in an elongated breathing tube. The pupae, or tumblers, are also aquatic and active. Their heads and thoraxes are encompassed in a large, oval mass with the slender abdomen attached. The most notable species are yellow-fever mosquito (*A. aegypti*), Asian tiger mosquito (*A. albopictus*), saltmarsh mosquito (*A. sollicitans*), and inland floodwater mosquito (*A. vexans*). The adults of these species are described below:

The yellow-fever mosquito has silvery-white or yellow-white stripes on its black body and a pattern of stripes on the top of its thorax (segments with legs and wings attached) which looks very much like a tuning fork. The last five segments of the legs are also banded in white.

The Asian tiger mosquito has silvery white markings on its black body, the top of the thorax has a single silver-white stripe down the center and silver-white bands around the abdominal segments. The last five segments of the legs are banded in white.

The salt marsh mosquito has white bands on the last leg segments and a band across the middle of the mouthparts. The abdomen has bands on most of the segments and a white stripe down the middle of the top of the abdomen.

The inland floodwater mosquito is medium-sized and brown with a band of white scales at the end of each abdominal segment. Each band is notched with a dark "V" that appears to cut it into two pieces. It has very narrow, creamy white-to-bronze bands on the last five segments of the legs.

Biology: The eggs, which are laid singly above the water line, hatch when the water rises and contacts the eggs. In the absence of sufficient water, the eggs remain dormant for many years. Mosquitoes undergo complete metamorphosis, i.e., egg, larva, pupa, and adult. The larva and pupa require water for development. Under ideal conditions, these mosquitoes complete their life cycles in ten days.

Habits: Asian tiger and yellow-fever mosquito larvae are found in artificial containers, such as discarded tin cans, tires, flower containers in cemeteries, etc. Both species are daytime biters, although the yellow-fever mosquito also bites at dawn and dusk as do most other species of mosquitoes. The salt marsh mosquito develops in salt marshes, and is found in coastal areas from New Jersey to Texas. This species can travel up to 20 miles from the breeding site. The inland floodwater mosquito is found throughout the United States and develops in flood waters, grassland pools and irrigated pastures.

In the summer, mosquitoes become major nuisances as they seek out your customers for their blood meal. Most service calls result when a customer plans activities in the backyard, such as an evening party, or there is a problem at recreational sites, such as golf courses. Perennial problems result from mosquito breeding sites located in inaccessible areas off the customer's premises, such as salt marshes, swamps, storage sites, and in other sites which are difficult to manage, such as city sewers and under structures.

Control: It is important to identify the species which is causing the problem as well as its breeding site(s). The most common methods of collecting mosquitoes for identification are use of light traps, siphoning tubes at resting sites, and dippers to collect larvae. Light traps can be small and portable (CDC) or large and permanently wired (New Jersey).

Both styles are designed to catch specimens without destroying them on sticky boards or by electrocution. The effectiveness of night-operated traps can be enhanced by using dry ice or CO_2. Adults can be collected by using a siphoning jar trap. The best areas for collection during the day are in shady areas, such as inside and on the shady side of sheds and buildings, sewers, culverts, drainage pipes, etc. Collection of larvae is more labor-intensive because it involves investigation of potential breeding sites which may not be accessible and sampling the site in several locations using a dipper.

Lighting around the property can result in problems with mosquitoes. Because lights are beacons for mosquitoes, attraction to the area should be reduced by locating lights away from the structure and directing the mosquitoes toward them. Sodium vapor and/or yellow bug lights should be used instead of incandescent or mercury vapor lights.

Regardless of the species, the key to mosquito control is source reduction, i.e., the removal of the water breeding site. For container breeders this is usually resolved by site clean-up. However, flood water mosquito problems usually require the use of a larvacide. The two most popular larvacides are the insect growth regulator (IGR) methoprene and the bacteria *Bti*. Liquid, granules, and extended release formulations are available to treat the site even when water is not present. When the product is exposed to water, the active ingredient is released into the water and kills the larvae.

Adulticiding with aerosols and ultra low volume products is effective in killing the mosquitoes within the immediate area but does not provide long term control. Residual spraying of mosquito resting sites also reduces the problem. The use of repellents should be considered. Skin repellents, such as DEET, and clothing repellents which contain permethrin are very effective. At their best, Tiki torches, citronella candles, and similar products provide temporary relief. Mosquito problems require considerable customer education because of the diversity of species, their often-remote breeding sites, and the ability of many species to travel long distances.

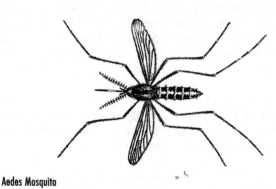

Aedes Mosquito

ANOPHELES MOSQUITOES

Order/Family:
Diptera/Culicidae

Scientific Name:
Anopheles spp.

Description: Mosquitoes are members of the order Diptera, i.e., flies. Mosquitoes are slender, long-legged, two-winged insects that are 1/8- to 1/4-inch long. They are unlike other flies as their wings and bodies are covered with scales and they have long piercing and sucking mouthparts. The *Anopheles* mosquitoes have no scales on their abdomens as do *Aedes* or *Culex* mosquitoes. They are easily recognized by their mouthparts (i.e., two palps and proboscis) which are equal in length as well as their habit of resting in a straight line at a 45⁰ angle to the surface.

The larvae, or wrigglers, are aquatic with well developed heads, swollen and unsegmented thoraxes, and eight-segmented abdomens which usually end in an elongated breathing tube. *Anopheles* larvae have no elongated breathing tube as do the *Aedes* or *Culex* mosquitoes, and they orient themselves parallel to the surface of the water when they are resting or breathing. The pupae, or tumblers, are also aquatic and active. Their heads and thoraxes are encompassed in a large, oval mass with the slender abdomen attached.

Anopheles quadramaculatus and *A. freeborni* are the two species known as malaria mosquitoes; another common species of this genus is *A. punctipennis.* The adults of these species are described below:

Both species of malaria mosquitoes have similar characteristics. They are dark brown, have unbanded legs, and four patches of dark scales on each wing.

A. punctipennis are medium-sized, brown mosquitoes with hairy, second body segments which are white in the middle and dark brown on the sides. The second and third leg segments are tipped in white. The wings have patches of light scales and yellow and orange markings.

Biology: The adults overwinter in sheltered locations, and in the spring the females seek blood meals, after which they stay secluded until they are ready to lay their eggs. They lay eggs, which have tiny floats attached to them, singly on the surface of clean, unpolluted water. The eggs hatch in two- to-six days with total development taking about two weeks. Under ideal conditions, the life cycle of *A. punctipennis* requires 24 days.

Habits: *Anopheles quadrimaculatus* is common in the eastern United States. *A. freeborni* is found west of the Rocky Mountains. The adult mosquitoes of both species are capable of transmitting malaria. They breed in ditches, ponds, swamps, ditches, and puddles. These mosquitoes are problems in areas where extensive irrigation of crops occurs. They enter structures to feed on humans, but, although aggressive, they are not painful biters.

A. *punctipennis* mosquitoes are found throughout the United States except in hot, dry areas. They breed in slow-moving streams and temporary pools preferring clean, partly-shaded water. The adults are outdoor feeders that seldom enter structures.

In the summer, mosquitoes become major nuisances as they seek out humans for blood meals. Most service calls result when a customer plans activities in the backyard, such as an evening party, or there is a problem at recreational sites, such as golf courses. Perennial problems result from mosquito breeding sites located in inaccessible areas off the customer's premises, such as salt marshes, swamps, storage sites, and in other difficult to manage sites, such as city sewers and under structures.

Control: It is important to identify the species which is causing the problem as well as the breeding site(s). The most common methods of collecting mosquitoes for identification are use of light traps, siphoning tubes at resting sites, and dippers to collect larvae. Light traps can be small and portable (CDC) or large and permanently wired (New Jersey). Both styles are designed to catch specimens without destroying them on sticky boards or by electrocution. The effectiveness of night-operated traps can be enhanced by using dry ice or CO_2. Adults can be collected by using a siphoning jar trap. The best areas for collection during the day are in shady areas, such as inside and on the shady side of sheds and buildings, sewers, culverts, drainage pipes, etc.

Collection of larvae is more labor-intensive because it involves investigation of potential breeding sites which may not be accessible as well as sampling the site in several locations using a dipper.

Lighting around the property can result in mosquito problems. Because lights are beacons for mosquitoes, the attraction of the area should be reduced by locating lights away from the structure and directing the mosquitoes toward them. Sodium vapor and/or yellow bug lights, should be used rather than incandescent or mercury vapor lights.

Regardless of the species, the key to mosquito control is source reduction, i.e., the removal of the water breeding site. However, flood water mosquito problems usually require the use of a larvacide. The two most popular larvacides are the insect growth regulator (IGR) methoprene and the bacteria *Bti*. Liquid, granules, and extended-release formulations are available to treat the site even when water is not present. When the product is exposed to water, the active ingredient is released into the water and kills the larvae.

Adulticiding with aerosols and ultra low volume products is effective in killing the mosquitoes within the immediate area but does not provide long-term control. Residual spraying of mosquito resting sites also reduces the problem.

The use of repellents should be considered. Skin repellents, such as DEET, and clothing repellents which contain permethrin are very effective. At their best, Tiki torches, citronella candles, or similar products offer temporary relief.

Mosquito problems require considerable customer education because of the diversity of species, their often-remote breeding sites, and the ability of many species to travel long distances.

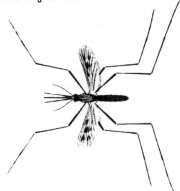

Anopheles Mosquito

CULEX MOSQUITOES

Order/Family:
Diptera/Culicidae

Scientific Name:
Culex spp.

Description: Mosquitoes are members of the order Diptera, i.e., flies. Mosquitoes are slender, long-legged, two-winged insects that are 1/8- to 1/4-inch long. They are unlike other flies as their wings and bodies are covered with scales, and they have long piercing and sucking mouthparts. *Culex* mosquitoes have blunt abdomens, in contrast to the pointed abdomens of *Aedes* mosquitoes.

The larvae, or wrigglers, are aquatic with well-developed heads, swollen and unsegmented thoraxes, and eight-segmented abdomens which end in an elongated breathing tube. The pupae, or tumblers, are also aquatic and active. Their heads and thoraxes are encompassed in a large, oval mass with the slender abdomen attached.

The most notable species are the northern and southern house mosquitoes, *Culex pipiens* and *C. quinquefasciatus*, respectively, and the encephalitis mosquito, *C. tarsalis*. The adults of these species are described below:

Northern and southern house mosquitoes are virtually identical in appearance. They are medium-sized, dull brown to brown-yellow, with even white bands at the end of each abdominal segment.

The encephalitis mosquito is a medium-sized dark mosquito with a white band across the middle of the proboscis (i.e., mouthparts). This characteristic separates it from almost any other *Culex* mosquito found in the United States.

Biology: Female southern and northern house mosquitoes lay their eggs in rafts which contain from 50 to 200 eggs on the water surface. They hatch in one or two days and complete development in from ten days to two weeks. Encephalitis mosquito adults overwinter in protected locations, and, after mating, lay their eggs in rafts which contain from 200 to 300 eggs on the surface of water. The eggs, which are laid in all types of bodies of water, hatch in two to three days, and, under ideal conditions, complete development within 18 days.

Habits: Southern and northern house mosquitoes prefer to breed in water with high organic matter content and that is contained in artificial containers such as cans, rain barrels, storm sewer basins, and in ponds and wells. The adults readily enter structures and are vicious biters which feed only at night. They rest within structures during the day. They also feed extensively on birds and are the major vector of St. Louis encephalitis.

The encephalitis mosquito is a common pest west of the Mississippi River, throughout Mexico, and in Canada. It breeds in ponds, canals, borrow ditches, mud puddles, cess pools, barrels, bird baths, and catchment basins. These mosquitoes enter buildings in search of a blood meal. Although they feed on humans and livestock, they prefer to feed on birds. They are an important carrier of western equine encephalitis, a disease of the nervous system.

In the summer, mosquitoes become major nuisances as they seek out humans for their blood meals. Most service calls result when a customer plans activities in the backyard, such as an evening party, or at recreational sites, such as golf courses. Perennial problems result from mosquito breeding sites located in inaccessible areas off the customer's premises, such as salt marshes, swamps, and storage sites and in other sites which are difficult to manage, such as city sewers and under structures.

Control: It is important to identify the species which is causing the problem as well as the breeding site(s). Common methods of collecting mosquitoes for identification are use of light traps, siphoning tubes at resting sites, and dippers to collect larvae. Light traps can be small and portable (CDC) or large and permanently wired (New Jersey). Both styles are designed to catch specimens without destroying them on sticky boards or by electrocution. The effectiveness of night-operated traps can be enhanced by using dry ice or CO_2. Adults can be collected by using a siphoning jar trap. The best areas for collection during the day are in shady areas, such as inside or on the shady side of sheds and buildings, sewers, culverts, and drainage pipes.

The collection of larvae is more labor-intensive because it involves investigating potential breeding sites, which may not be accessible, as well as sampling the site in several locations using a dipper.

Lighting around the property can result in problems with mosquitoes. Because lights are beacons for mosquitoes, the attractiveness of the area should be reduced by locating lights away from the structure and directing them toward it. Sodium vapor and/or yellow bug lights, should be used rather than incandescent or mercury vapor lights.

Regardless of the species, the key to mosquito control is source reduction, i.e., the removal of the water breeding site. For container breeders, this is usually resolved by site clean-up. However, flood water mosquito problems usually require the use of a larvacide. The two most popular larvacides are the insect growth regulator (IGR) methoprene and the bacteria *Bti*. Liquid, granules, and extended release formulations are available to treat the site even when water is not present. When the product is exposed to water the active ingredient is released into the water and kills the larvae.

Adulticiding with aerosols and ultra low-volume products is effective in killing the mosquitoes in the immediate area but does not provide long term control. Residual spraying of mosquito resting sites also reduces the problem.

The use of repellents should be considered. Skin repellents, such as DEET, and clothing repellents which contain permethrin are very effective. At their best, Tiki torches, citronella candles, and similar products offer temporary relief. Mosquito problems require considerable customer education because of the diversity of species, their often remote breeding sites, and the ability of many species to travel long distances.

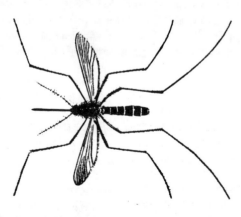

Culex Mosquito

PSOROPHORA MOSQUITOES⁻

Order/Family:
Diptera/Culicidae

Scientific Name:
Psorophora cofinnis (Lynch-Arribalzaga)

Description: Mosquitoes are members of the order Diptera, i.e., flies. Mosquitoes are slender, long-legged, two-winged insects that are 1/8- to 1/4-inch long. They are unlike other flies as their wings and bodies are covered with scales, and they have long piercing and sucking mouthparts.

The larvae, or wrigglers, are aquatic and have well-developed heads, swollen and unsegmented thoraxes, and eight-segmented abdomens which end in an elongated breathing tube. The pupae, or tumblers, are also aquatic and active. Their heads and thoraxes are encompassed in a large, oval mass with the slender abdomen attached.

The dark rice field mosquito is a medium-to-large mosquito that is typically dark-colored with white speckled wings. The proboscis (i.e., mouthparts), last five leg segments, and abdomen are banded in white. The bands on the second and third abdominal segments are triangular.

Biology: Females lay their eggs on damp soil in areas that are infrequently flooded. If submerged immediately after having been laid, the eggs hatch within four or five days. When they are allowed to dry for two to three weeks, they will hatch almost immediately when submerged. The larvae complete development in about four days, pupate for two days, and emerge as adults.

Habits: Dark rice field mosquitoes produce great swarms of adults when conditions for their development is optimal. These vicious biters feed on humans and other animals night and day and are especially a problem in areas where crops are irrigated. They are attracted to lights and can make life very uncomfortable for people who work and play outdoors.

In the summer mosquitoes become major nuisances as they seek out humans for their blood meals. Most service calls result when a customer plans activities in the backyard, such as an evening party, or recreational sites, such as golf courses. Perennial problems result from mosquito breeding sites which are located in inaccessible areas off the customer's premises, such as salt marshes, swamps, storage sites, etc., and in other difficult to manage sites, such as city sewers, and under structures.

Control: It is important to identify the species that is causing the problem as well as their breeding site(s). The most common methods of collecting mosquitoes for the purpose of identification include the use of light traps, siphoning tubes at resting sites, and dippers to collect larvae. Light traps can be small and portable (CDC) or large and permanently wired (New Jersey). Both styles are designed to catch specimens without destroying them on sticky boards or by electrocution. The effectiveness of night-operated traps can be enhanced by using dry ice or CO_2. Adults can be collected by using a siphoning jar trap. The best areas for collection during the day are in shady areas, such as inside or on the shady side of sheds and buildings, sewers, culverts, and drainage pipes.

Collection of larvae is more labor-intensive as it involves the investigation of potential breeding sites which may be inaccessible as well as sampling the site in several locations using a dipper.

Lighting around the property can result in mosquito problems. Because lights are beacons for mosquitoes, it is helpful to reduce the attraction of the area by locating lights away from the structure and directing the mosquitoes toward them. Sodium vapor and/or yellow bug lights, should be used rather than incandescent or mercury vapor lights.

Regardless of the species, the key to mosquito control is source reduction, i.e., the removal of the water breeding site. For container breeders this is usually resolved by site clean-up. However, flood water mosquito problems generally require the use of a larvacide.

The two most popular larvacides are the insect growth regulator (IGR) methoprene and the bacteria *Bti*. Liquid, granules, and extended release formulations are available to treat the site even when water is not present. When the product is exposed to water, the active ingredient is released into the water and kills the larvae.

Adulticiding with aerosols and ultra low volume products is effective in killing the mosquitoes in the immediate area but does not provide long term control. Residual spraying of mosquito resting sites can also reduce the problem.

The use of repellents should be considered. Skin repellents, such as DEET, and clothing repellents which contain permethrin are very effective. At their best, Tiki torches, citronella candles, and similar products provide temporary relief.

Mosquito problems require considerable customer education because of the diversity of species, their often-remote breeding sites, and the ability of many species to travel long distances.

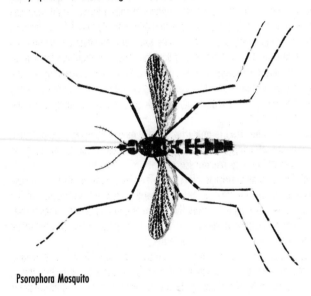

Psorophora Mosquito

AMERICAN COCKROACH

Order/Family:
Blattodea/Blattidae

Scientific Name:
Periplaneta americana (Linnaeus)

Description: American cockroaches are 1-3/8 to 2-1/8-inches long when mature, red-brown, and characterized by fully-developed wings that completely cover the abdomen. The pronotum (i.e., shield like area behind the head) has a dirty-yellow band around its edge.

The nymphs are 1/4-inch long when they emerge from the egg capsule and initially are gray-brown. As they develop, they become more red-brown, and the yellow band becomes more prominent on the pronotum. The purse-shaped egg capsule (i.e., ootheca) is dark red-brown in color, 3/8-inch long, and typically has eight eggs per side.

Biology: During her lifetime, the female American cockroach produces from nine to ten egg capsules each of which contains from 14 to 16 eggs. The capsules are dropped or, using secretions from her mouth, glued in protected locations such as cracks and crevices near food sources. The nymphs molt from 10 to 13 times before becoming adults. This requires about 600 days. Adult females live an average of 440 days and males about 200 days. Large populations of American cockroaches accumulate in secluded locations because they live for such a long time.

Habits: American cockroaches are not common pests in most homes. They can be abundant in sewers and commercial facilities, e.g., groceries, prisons, restaurants, hospitals, and office and apartment buildings. They prefer to inhabit warm, damp locations, e.g., steam tunnels and boiler rooms. They are strong fliers and easily migrate from building to building. In the summer, large numbers accumulate in outdoor locations, e.g., in dumps, alleys, and yards, and in the fall, migrate into surrounding structures. Although they feed on a variety of materials, they prefer fermenting foods.

Control: Cockroaches often are brought into and moved between facilities via equipment and storage boxes. Thus, potentially infested products which are brought into structures should be closely inspected. Many types of cardboard and plastic sticky traps are available to help pinpoint sources of cockroach infestation and to monitor areas about which occupants have complained but infestations can not be visually detected. Sticky traps are not intended for control but, rather, to guide and evaluate control efforts as part of the inspection process. Visual inspections can be conducted using a flashlight and aerosol pyrethrin to flush cockroaches from their harborages.

An effective cockroach management program depends on good sanitation in order to eliminate the food, water and harborage they need for survival. It is critical to reduce clutter as large cockroaches like to hide in stacked boxes, cartons, rolled carpeting and stored paper and cardboard materials, especially in dark, damp locations. Vacuum cleaning can be used to physically remove exposed cockroaches.

American cockroaches are particularly sensitive to drying so it is important to reduce moisture by repairing leaks, improving drainage, and installing screened vents in order to increase airflow. Permanent reduction of cockroach populations can be achieved by caulking to eliminate harborage and prevent entry into structures.

The most important cracks to eliminate include those at which sinks and fixtures are mounted to the wall and/or floor, around all types of plumbing, baseboard molding and corner guards where shelves and cabinets meet walls and door frames, and cracks on or near food preparation surfaces. Cockroach access routes from wall voids into occupied spaces and around plumbing and electrical fixtures should be sealed with caulk or grout. Basement floor drains should be protected with screens or basket inserts which must be cleaned regularly.

Containerized, paste, dry, and gel baits have become very popular in the industry and are very effective products for eliminating cockroaches. To maximize effectiveness, paste, gel and dry baits should be applied with a syringe-like dispensing tool in many small dabs or spots that are close to harborage sites. Large, plastic bait containers should be placed as close as possible to the dark concealed spots where cockroaches are living, preferably adjacent to edges and corners.

Large populations of cockroaches can be reduced or eliminated with careful application of sprays and dusts. Many insecticides break down rapidly within the moist, hot locations these cockroaches prefer so an appropriate formulation for the environmental conditions must be selected.

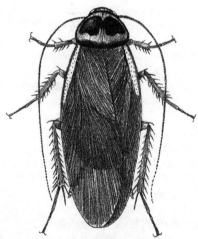

American Cockroach

PEST IDENTIFICATION GUIDE

Carpenter Ant
See page 114

Fire Ant
See page 118

Odorous House Ant
See page 122

Pavement Ant
See page 124

Bed Bug
See page 132

Cat Flea
See page 142

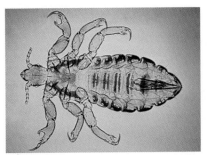

Body Louse
See page 148

Aedes Mosquito
See page 154

Culex Mosquito
See page 160

Brown Dog Tick
See page 138

Blacklegged Tick
See page 140

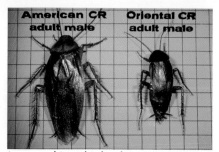

American and Oriental Cockroaches
See pages 166 and 178

Brown-Banded Cockroach
See page 169

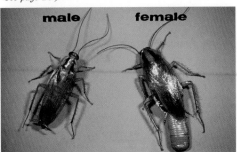

German Cockroach
See page 175

Oriental Cockroach
See page 178

Pennsylvania Wood Cockroach
See page 181

Smokybrown Cockroach
See page 183

Blow/Bottle Fly
See page 188

Cluster Fly
See page 190

Fruit Fly
See page 194

House Fly
See page 198

Boxelder Bug
See page 200

Earwig
See page 206

House Cricket
See page 212

Lady Bug
See page 216

Centipede
See page 202

Cigarette and Drugstore Beetles
See pages 226 and 230

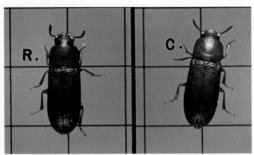

Red and Confused Flour Beetles
See page 236

Indian Meal Moth
See page 248

Saw-Toothed Grain Beetle
See page 238

Silverfish
See page 264

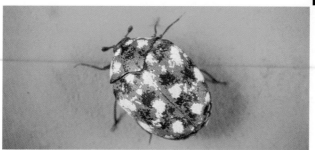

Varied Carpet Beetle
See page 242

Webbing Clothes Moth
See page 252

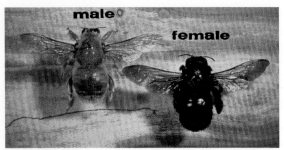

Carpenter Bee
See page 284

Drywood Termite
See page 276

Formosan Termite — Soldier
See page 278

Subterranean Termite
See page 281

House Spider
See page 288

Black Widow Spider
See page 290

Brown Recluse Spider
See page 292

Honey Bee
See page 296

Social Wasp — Yellow Jacket
See page 298

Solitary Wasp — Mud Dauber
See page 301

Bat
See page 306

House Sparrow
See page 308

House Mouse
See page 316

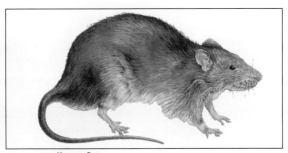

Norway Rat
See page 318

Roof Rat
See page 320

BROWN-BANDED COCKROACH

Order/Family:
Blattodea/Blattidae

Scientific Name:
Supella longipalpa (Fabricius)

Description: Brown-banded cockroaches are about 1/2-inch long when mature, light brown to brown, and have two light, yellow-brown bands running across their bodies; hence, their common name. The pronotum (i.e., shield-like segment behind the head) has a dark brown area which is shaped like a liberty bell. Females are darker in color and broader than males, and their wings cover only three-quarters of their abdomens. Wings of the male completely cover their abdomens. The colored bands are much easier to see on nymphs because they lack wings to obscure them. The purse-shaped egg capsule (i.e., ootheca) is light brown, approximately 1/4-inch long, slightly bowed, and typically has seven to nine eggs per side.

Biology: During her lifetime, the female brown-banded cockroach produces about 14 egg capsules each of which contains 14 to 18 eggs. The female carries the egg capsule for 24 to 36 hours, then, using secretions from her mouth, attaches it to protected areas, e.g., underside of shelves and furniture and inside televisions and other appliances. Nymphs emerge in about 70 days and molt six to eight times before becoming adults. This requires 90 to 276 days. The adults live about six months.

Habits: Brown-banded cockroaches prefer a warmer and drier environment than do German cockroaches. Thus, they are not nearly so common in houses. They are found throughout structures, preferring hiding places up off the floor, e.g., behind crown molding, pictures, tapestries and other wall hangings, and in closets, furniture, appliances, computers, and telephones.

Control: Established populations of brown-banded cockroaches are difficult to control because they are found throughout structures and not in areas commonly considered to be cockroach harborages. It is important to inspect thoroughly for these insects in order to find as many of their food sources and harborage sites as possible. Many types of cardboard and plastic sticky traps are available to help pinpoint sources of cockroach infestation and to monitor areas about which occupants have complained but infestations can not be visually detected. Sticky traps are not intended for control but, rather, to guide and evaluate control efforts as part of the inspection process. Visual inspections can be conducted using a flashlight and aerosol pyrethrin to flush cockroaches from their harborages.

An effective cockroach management program depends on good sanitation to eliminate the food, water, and harborage they need for survival. Cleanup to reduce cockroaches in the home and office environments must focus mainly on the food residue in and around coffee machines, microwave ovens, stoves, refrigerators, trash cans, furniture, and areas where exposed food is stored. In addition, it is critical to reduce clutter as cockroaches like to hide in stacked boxes, cartons, and stored paper and cardboard materials, especially in dark, damp locations near food.

Vacuum cleaning can be used to physically remove exposed cockroaches. Infested equipment may have to be disassembled and/or placed in large plastic bags and treated with carbon dioxide, heat, cold, or simply held in the bag with containerized bait until the cockroaches die. The choice of treatment will depend on its potential effects on the equipment.

Permanent reduction of cockroach populations can be achieved by caulking to eliminate harborage. The most important cracks to eliminate include those at which sinks and fixtures are mounted to the wall, around crown molding and corner guards, and where shelves and cabinets meet walls and door frames.

Containerized, paste, dry, and gel baits have become very popular within the industry and are very effective products for eliminating cockroaches. To maximize effectiveness, paste, gel and dry baits should be applied with a syringe-like dispensing tool in many small dabs or spots that are close to harborage sites. Small, plastic bait containers should be placed as close as possible to the dark concealed spots where cockroaches are living, preferably adjacent to edges and corners.

Large populations of cockroaches can be reduced or eliminated with careful application of sprays and dusts. However, these products can not be applied inside electrical appliances or computers. Many insecticides break down rapidly so an appropriate formulation for the environmental conditions must be selected.

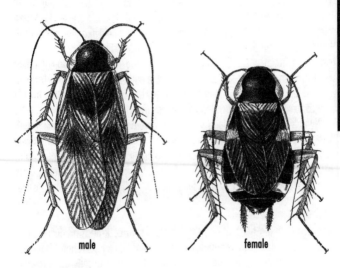

male   female

Brown-Banded Cockroach

BROWN COCKROACH

Order/Family:
Blattodea/Blattidae

Scientific Name:
Periplaneta brunnea Burmeister

Description: Brown cockroaches are 1-1/4 to 1-1/2-inches long when mature, red-brown, and characterized by fully developed wings that completely cover their abdomens. The pronotum (i.e., shield like area behind the head) has a faint dirty-yellow band around its edge. Brown cockroaches look similar to American cockroaches except that the last segment on the two cerci (i.e., the antenna-like projections at the end of the abdomen) is triangular and less than twice as long as it is wide. The last segment on the American cockroach cerci is much longer and narrower.

Nymphs are 1/4-inch long when they emerge from the egg capsule and initially are brown-to-dark brown in color. As they develop they become more red-brown and pale markings appear on their thoraxes (i.e., the three segments behind their heads). The purse-shaped egg capsule (i.e., ootheca) is almost black in color, 1/2-inch long, and typically has 12 to 14 eggs per side.

Biology: During her lifetime, the female brown cockroach produces up to 32 egg capsules each contains 21 to 28 eggs. Using secretions from her mouth, the female glues the capsules near the ceiling usually on plaster or concrete. The capsule is covered with materials from the surrounding area. Nymphs emerge in about 35 days and develop over an average of 182 days before becoming adults which live about 244 days.

Habits: This cockroach has not been able to become successfully established within northern states but is common in southern states as far west as Texas. It is not a common pest in houses but can be abundant in sewers and commercial facilities, e.g., groceries, prisons, restaurants, hospitals, and office and apartment buildings. It prefers to inhabit warm, damp locations, e.g., steam tunnels and boiler rooms. In the summer, large numbers accumulate in outdoor locations, e.g., in leaf litter, ground covers, trees (especially palms), and dumps. In the fall, they migrate into surrounding structures.

Control: Cockroaches often are brought into and moved between facilities via equipment and storage boxes. Thus, potentially infested products which are brought into structures should be closely inspected. Many types of cardboard and plastic sticky traps are available to help pinpoint sources of cockroach infestation and to monitor areas about which occupants have complained but infestations can not be visually detected. Sticky traps are not intended for control but, rather, to guide and evaluate control efforts as part of the inspection process. Visual inspections can be conducted using a flashlight and aerosol pyrethrin to flush cockroaches from their harborages.

An effective cockroach management program depends on good sanitation to eliminate the food, water, and harborage they need for survival. In addition, it is critical to reduce clutter as large cockroaches like to hide in stacked boxes, cartons, rolled carpeting, stored paper, and cardboard materials, especially in dark, damp locations. Vacuum cleaning can be used to physically remove exposed cockroaches. Brown cockroaches are particularly sensitive to drying so it is important to reduce moisture by repairing leaks, improving drainage, and installing screened vents in order to increase airflow.

Permanent reduction of cockroach populations can be achieved by caulking to eliminate harborage and prevent entry into structures. The most important cracks to eliminate include those at which sinks and fixtures are mounted to the wall and/or floor, around all types of plumbing, baseboard molding and corner guards where shelves and cabinets meet walls and door frames, and cracks on or near food preparation surfaces. Cockroach access routes from wall voids into occupied spaces and around plumbing and electrical fixtures should sealed with caulk or grout. Basement floor drains should be protected with screens or basket inserts which must be cleaned regularly.

Scatter, containerized, paste, dry, and gel baits have become very popular within the industry and are very effective products for eliminating cockroaches. To maximize effectiveness, paste, gel and dry baits should be applied with a syringe-like dispensing tool in many small dabs or spots that are close to harborage sites. Large, plastic bait containers should be placed as close as possible to the dark concealed spots where cockroaches are living, preferably adjacent to edges and corners. Scatter baits are particularly effective outdoors and in crawl spaces.

Large populations of cockroaches can be reduced or eliminated with careful application of sprays and dusts. Many insecticides break down rapidly within the moist, hot locations these cockroaches prefer so an appropriate formulation for the environmental conditions must be selected.

Brown Cockroach

GERMAN COCKROACH

Order/Family:
Blattodea/Blattellidae

Scientific Name:
Blattella germanica (Linnaeus)

Description: German cockroaches are 1/2- to 5/8-inches long when mature, light brown to tan, and have fully developed wings. The pronotum (i.e., shield-like segment behind the head) has two dark parallel bars on it. The adult males are somewhat narrower than the females when viewed from below. The nymphs, 1/8-inch long when they emerge from the egg capsule, are almost uniformly dark except for a light tan area on the back of the second and third segments. As they develop, the light tan area becomes larger until, as mature nymphs, they have two parallel black bars separated by a light tan area. The purse-shaped egg capsule of the German cockroach (i.e., ootheca) is light brown in color, 1/4- to 3/8-inch long, and typically has 15 to 20 eggs per side.

Biology: During her lifetime, the female German cockroach produces four to eight egg capsules each of which contains 30 to 40 eggs. The female carries the egg capsule partially within her abdomen until just before the nymphs are ready to emerge. Approximately one to two days before hatching, she drops the egg capsule in a protected area. If the egg capsule is dropped prematurely, the developing roaches inside die of dehydration. Nymphs molt six to seven times before becoming adults. This requires about 103 days; thus allowing three to four generations per year. Adults live 100 to 200 days. Established German cockroach populations consist of approximately 75% nymphs.

Habits: German cockroaches are the most common household insect within the United States. This pest typically infests kitchens and bathrooms but will live anywhere inside heated structures in which there is food, water, and harborage. They rarely are found outdoors and then only during warm weather. German cockroaches gain entry into structures in grocery bags, cardboard boxes, drink cartons, infested equipment such as used refrigerators, toasters, microwaves, etc. Cockroaches feed on all types of human food, as well as on pet food, toothpaste, soap, glue, etc.

German cockroaches are active at night, leaving their harborage to find food and water. They remain hidden in dark, secluded harborage areas, e.g., under cupboards, behind cabinets, in wall voids, and around motor housings in appliances where they spend 75% of their time. At most, only one third of the population forages at night. Observation of foraging cockroaches during the day is a good indication that there is a tremendous population. Cockroaches congregate in harborage sites; but as the population increases, overcrowding forces some of them to relocate.

Control: Because German cockroaches typically are brought into structures, potentially infested products should be closely inspected. Many types of cardboard and plastic sticky traps are available to help pinpoint sources of cockroach infestation and to monitor areas about which occupants have complained but infestations can not be visually detected. Sticky traps are not intended for control but, rather, to guide and evaluate control efforts as part of the inspection process. Visual inspections can be conducted using a flashlight and aerosol pyrethrin to flush cockroaches from their harborages.

An effective cockroach management program depends on good sanitation to eliminate the food, water, and harborage they need for survival. Cleanup to reduce cockroaches in the home and office environment must focus mainly on the food residue in and around coffee machines, microwave ovens, stoves, refrigerators, trash cans, furniture, and areas where exposed food is stored. It is critical to reduce clutter as cockroaches like to hide in stacked boxes, cartons, and stored paper and cardboard materials, especially in dark, damp locations, near food. Vacuum cleaning can be used to physically remove exposed cockroaches.

Permanent reduction of cockroach populations can be achieved by caulking to eliminate harborage. The most important cracks to eliminate include those at which sinks and fixtures are mounted to the wall and/or floor, around all types of plumbing, baseboard molding and corner guards where shelves and cabinets meet walls and door frames, and cracks on or near food preparation surfaces. Cockroach access routes between apartments and from wall voids, and around plumbing and electircal fixtures should be sealed with caulk or grout. Basement floor drains should be protected with screens or basket inserts which should be cleaned regularly.

Containerized, paste, dry, and gel baits have become very popular within the industry and are very effective products for eliminating cockroaches. To maximize effectiveness, paste, gel, and dry baits should be applied with a syringe-like dispensing tool in many small dabs or spots that are close to harborage sites. Small, plastic bait containers should be placed as close as possible to the dark concealed spots where cockroaches are actually living, preferably adjacent to edges and corners.

Large populations of cockroaches can be reduced or eliminated with careful application of sprays and dusts. Many insecticides break down rapidly within the moist, hot locations these cockroaches prefer so an appropriate formulation for the environmental conditions must be selected.

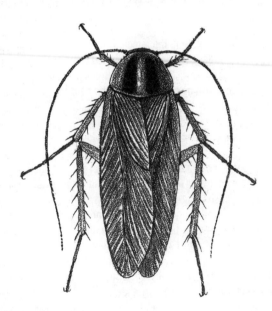

German Cockroach

ORIENTAL COCKROACH

Order/Family:
Blattodea/Blattidae

Scientific Name:
Blatta orientalis Linnaeus

Description: Male Oriental cockroaches are 1-inch long and females are 1-1/4-inches long when mature, and red brown to black. In males, the wings cover 75% of the abdomen; in females, they are reduced to small wing pads.

The early nymphs are light brown but become increasingly darker with each molt. The purse-shaped egg capsule (i.e., ootheca) is dark red-brown but becomes black with age, 3/8-inch long, and typically has eight eggs per side.

Biology: During her lifetime the female Oriental cockroach produces an average of eight egg capsules each of which contains 16 eggs. The capsules are dropped or, using secretions from her mouth, glued in protected locations such as cracks and crevices near food sources. The nymphs molt about 10 times before becoming adults. Depending on temperature, this requires 206 to 800 days. Adult females live from 34 to 181 days and males from 112 to 160 days.

Habits: Oriental cockroaches are not common pests in most homes. They can be abundant, however, in sewers and commercial facilities, e.g., groceries, prisons, restaurants, hospitals, office and apartment buildings. Indoors they can become very abundant in damp, secluded places such as crawl-spaces, basements, water meter boxes, and drains.

They often are found in bathtubs and sinks because they lack the small pads on their tarsi (i.e., last segments of the legs) commonly found on other cockroaches, which allow them to crawl up smooth surfaces. Outdoors, even in cold weather, they are found in planters, ground covers, stones, leaf litter, and other debris. Oriental cockroaches prefer to feed on starchy foods but eat various other items including decaying organic matter. Oriental cockroaches produce a very characteristic pungent "cockroach" odor.

Control: Many types of cardboard and plastic sticky traps are available to help pinpoint sources of cockroach infestation and to monitor areas about which occupants have complained but infestations can not be visually detected. Sticky traps are not intended for control but, rather, to guide and evaluate control efforts as part of the inspection process. Visual inspections can be conducted using a flashlight and aerosol pyrethrin to flush cockroaches from their harborages.

An effective cockroach management program depends on good sanitation to eliminate the food, water, and harborage they need for survival. It is critical to reduce clutter as large cockroaches like to hide in stacked boxes, cartons, rolled carpeting, and stored paper and cardboard materials, especially in dark, damp locations such as basements and crawlspaces. Vacuum cleaning can be used to physically remove exposed cockroaches.

Permanent reduction of cockroach populations can be achieved by caulking to eliminate harborage and prevent entry into structures. The most important cracks to eliminate include those at which sinks and fixtures are mounted to the wall and/or floor, around all types of plumbing, and cracks on or near food preparation surfaces. Cockroach access routes from wall voids into occupied spaces and around plumbing and electrical fixtures should be sealed with caulk or grout. Basement floor drains should be protected with screens or basket inserts which should be cleaned regularly.

Scatter, containerized, paste, dry, and gel baits have become very popular within the industry and are very effective products for eliminating cockroaches. To maximize effectiveness, paste, gel and dry baits should be applied with a syringe-like dispensing tool in many small dabs or spots that are close to harborage sites. Large, plastic bait containers should be placed as close as possible to the dark concealed spots where cockroaches are living, preferably adjacent to edges and corners. Scatter baits should be applied outdoors near harborage areas and in some indoor locations.

Large populations of cockroaches can be reduced or eliminated with careful application of sprays and dusts. Many insecticides break down rapidly within the moist, cool locations these cockroaches prefer so an appropriate formulation for the environmental conditions should be selected. Exterior barrier treatments with microencapsulated or wettable powder formulations might be effective.

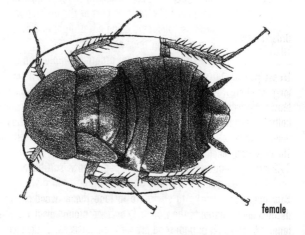

female

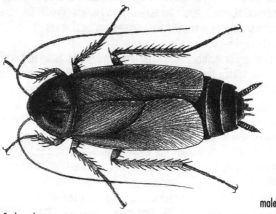

male

Oriental Cockroaches

PENNSYLVANIA WOOD COCKROACH

Order/Family:
Blattodea/Blattellidae

Scientific Name:
Parcoblatta pennsylvanica (De Geer)

Description: Male Pennsylvania wood cockroaches are 7/8- to 1-1/4-inches long, and females are 1/2- to 3/4-inch long when mature; both are chestnut brown. Males have fully developed wings; the wings of females cover about half of their abdomens. The pronotum (i.e., shield-like segment behind the head) is edged in creamy white. Nymphs are usually red-brown to dark brown. The purse-shaped egg capsule (i.e., ootheca) is yellow-brown, approximately 1/2-inch long, strongly bowed, and typically has 16 to 18 eggs per side.

Biology: During her lifetime, the female Pennsylvania wood cockroach produces about 30 egg capsules each of which contains about 32 eggs. The female deposits the egg capsule in protected areas, e.g., under the loose bark of dead trees, logs, stumps, etc. Nymphs hatch during the summer and overwinter in protected sites, maturing in the spring when the adults mate. Development time from egg to adult averages 318 days.

Habits: Adult males are strong fliers. They are attracted to lights, gaining entry to structures through cracks and gaps near the light source. Females can not fly and do not congregate around light but can crawl into structures through exterior openings. Adults and nymphs frequently are brought indoors on firewood. Wood cockroaches are more of a problem in structures located near woods.

Control: Control is seldom warranted as these cockroaches do not survive within structures. The occasional wood cockroach found within a structure can be removed with a vacuum; there is no need to apply a product indoors. A preventive strategy including caulking or otherwise sealing all exterior cracks and gaps, repairing screens and ensuring that they fit tightly, installing door sweeps on exterior doors, and screening vents, is the best course of action.

Because lights are beacons for male wood cockroaches, lighting around the property can result in problems; thus, the attractiveness of the area should be reduced by locating lights away from structures, directing wood roaches toward them. Sodium vapor and/or yellow bug lights, rather than incandescent or mercury vapor lights, should be used.

Exterior harborage sites and potential entry points should be treated with a residual insecticide; microencapsulated and wettable powder formulations are the most effective. However, overapplication of these formulations can leave a residue on decorative shake siding and similar surfaces. Various scatter baits are available and can be used around wood piles, wooded areas, mulch, and areas away from the structure where wood cockroaches are located.

Customers who live near wooded areas that are experiencing problems with wood cockroaches should be informed wood cockroaches are not indoor species; they should expect, however, to see the roaches flying around lights from May until October, and that a few may enter structures but die shortly afterward.

Pennsylvania Wood Cockroach

SMOKYBROWN COCKROACH

Order/Family:
Blattodea/Blattidae

Scientific Name:
Periplaneta fuliginosa (Serville)

Description: Smokybrown cockroaches are 1 to 1-1/4 inches long when mature, uniformly dark brown-to-mahogany, and characterized by fully developed wings that completely cover their abdomens. The pronotum (i.e., shield like area behind the head) is black. Nymphs are red brown; the younger ones have white markings on their backs and the first four or five segments of their antennae are white at the tip. The purse-shaped egg capsule (i.e., ootheca) is dark brown-to-black, 3/8-inch long, and typically has 10 to 14 eggs per side.

Biology: During her lifetime, the female smokybrown cockroach produces approximately ten egg capsules each of which contains about 20 eggs. Using secretions from her mouth, the female firmly attaches the egg cases to some surface or object and covers them with surrounding materials. Nymphs emerge in 50 days and molt 10 to 12 times. Depending on temperature, development from egg to adult requires from 160 to 716 days. Adult females live an average of 218 days and males about 215 days.

Habits: Smokybrown cockroaches typically are outdoor pests commonly found in southern states as far west as central Texas. Outdoors they often are found in wood piles, flower planters, palm trees, water oaks, and vacant buildings. Indoors they seek warm, humid areas without air circulation, such as garages, attics, and crawlspaces. Smokybrown cockroaches have also been found established in northern states, typically in greenhouses. These cockroaches are strong fliers and are attracted to lights.

Control: Smokybrown cockroaches enter structures from outdoors. Many types of cardboard and plastic sticky traps are available to help pinpoint sources of cockroach infestation and to monitor areas about which occupants have complained but infestations can not be visually detected. Sticky traps are not intended for control but, rather, to guide and evaluate control efforts as part of the inspection process. Visual inspections can be conducted using a flashlight and aerosol pyrethrin to flush cockroaches from their harborages.

An effective cockroach management program depends on good sanitation to eliminate the food, water, and harborage they need for survival. It is critical to reduce clutter as large cockroaches like to hide in stacked boxes, cartons, rolled carpeting, stored paper and cardboard materials, especially in dark, damp locations. Vacuum cleaning can be used to physically remove exposed cockroaches.

Smokybrown cockroaches like warm, humid environments so it is important to reduce moisture by repairing leaks and installing screened vents in order to increase airflow. Permanent reduction of cockroach populations can be achieved by caulking in order to eliminate harborage and prevent entry into structures. The most important cracks to eliminate are those around chimneys, soffits, roof joints, fascia boards, gutters, plumbing, and other penetrations of exterior walls.

Gaps around doors and windows should be sealed, and all vents and other exterior openings should be screened. Exterior lighting should be changed to reduce the likelihood of attracting smokybrown cockroaches to the structure and entry areas.

Containerized, paste, dry, and gel baits have become very popular within the industry and are very effective products for eliminating cockroaches. To maximize effectiveness, paste, gel, and dry baits should be applied with a syringe-like dispensing tool in many small dabs or spots that are close to harborage sites. Large, plastic bait containers should be placed as close as possible to the dark concealed spots where cockroaches are actually living, preferably adjacent to edges and corners. Scatter baits applied in outdoor locations, attics, and crawlspaces are effective in reducing populations.

Large populations of cockroaches can be reduced or eliminated with careful application of microencapsulated or wettable powder formulations to potential entry points and in exterior areas where these cockroaches are found. Many insecticides break down rapidly within the moist, hot locations these cockroaches prefer so an appropriate formulation for the environmental conditions must be selected.

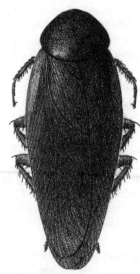

Smokybrown Cockroach

SURINAM COCKROACH

Order/Family:
Blattodea/Blaberidae

Scientific Name:
Pycnoscelus surinamensis (L.)

Description: Female Surinam cockroaches are 3/4- to 1-inch long, brown, and the pronotum (shield-like area behind the head) is dark brown- to-black. The wings are light brown and completely cover the abdomen. Males have never been found in the United States. Nymphs are dark brown-to-black and look much like immature Oriental cockroaches. The crescent-shaped egg capsule (i.e., ootheca) is light in color, soft, 1/2- to 5/8-inch long, and typically has 13 eggs per side.

Biology: The Surinam cockroach is parthenogenic (i.e., it produces egg capsules without mating with males). The egg capsules remain inside the female until the young nymphs emerge. Females produce three egg capsules each of which contains about 26 eggs. There is a 48 to 82 day interval between egg capsule production. Nymphs require 127 to 184 days to reach maturity, and development from egg to adult requires from 162 to 219 days. The female lives for approximately 307 days.

Habits: Surinam cockroaches are tropical insects usually found in humid and hot situations, e.g., outdoors in the Gulf Coast states from Florida to Texas. Isolated populations reported in greenhouses, mall and office atriums, and zoos throughout the United States have been introduced via potted plants shipped from tropical areas.

These cockroaches burrow into the soil to a depth of three to four inches where they construct a burrow containing nymphs and females incubating their egg capsules. They typically go unnoticed until large populations develop because they remain inside cracks and crevices and under leaf litter and mulch during the day, coming out at night to feed on plants. Surinam cockroaches are plant feeders and can severely damage plants in greenhouses and atriums.

Control: Removing harborage sites, e.g., leaf litter, mulch, landscape timbers, stones, and other objects close to a foundation is the key to a successful control program. When indoor infestations occur in commercial facilities, the plant supplier should be contacted as plants usually are the source of the infestation.

Using soil drenches that contain a microencapsulated or wettable powder formulation will successfully control Surinam cockroaches both outdoors and in indoor plantings and/or potted plants. Granular soil treatment formulations and baits are also available but require more time to achieve control. Before making a pesticide application, as much mulch as possible should be removed, then replaced with new mulch. During insecticide applications, care must be taken to protect fish, birds, and other animals in and around the infested area in atriums and zoos.

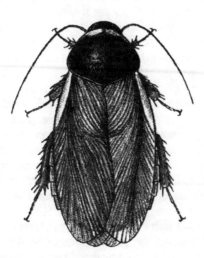

Surinam Cockroach

BLOW/BOTTLE FLIES

Order/Family:
Diptera/Calliphoridae

Scientific Name:
Phormia regina (Meigen) — Black Blow Fly
Phaenicia sericata (Meigen) — Green Bottle Fly
Calliphora spp. — Blue Bottle Flies

Description: Several different species of blowflies may infrequently infest structures. As a group they are easily identified by the metallic blue, green, or yellow-brown sheen of their stout bodies. They range in size from 1/4-inch for the black blow fly to more than 1/2-inch for the blue bottle flies. These flies are very active and usually are seen buzzing at windows. The black blowfly is black with a bluish-green lustre; the green bottle flies are metallic blue-green; and the blue bottle flies are metallic blue. These flies have sponging mouthparts and one pair of wings. Mature larvae (maggots) are 3/8- to 7/8-inch long, legless, eyeless, and taper from the larger round rear segment to the pointed head which consists of a pair of mouth hooks. They are cream-colored and have large spiracles on the posterior plate.

Biology: Depending on the species, females lay from 540 to 2,373 eggs in their lifetime. Eggs are laid in batches of 100 to 180 on meat, fish or carrion, but they are also attracted to animal manure, garbage, and rotting vegetable matter. The eggs hatch in less than a day. The larvae develop rapidly in this environment, molting three times in three to four days. The larvae typically leave the larval food to pupate in the soil for six to seven days. The larva and the pupa are the overwintering stages. Development time (egg to adult) can be as short as 10 days, but typically is 15 to 20 days.

Habits: Blow flies seldom are a significant problem in structures but can be quite annoying because of their persistent buzzing. One of the first signs of an infestation is when the larvae leave the breeding medium to pupate. These flies frequently are found developing in the decaying bodies of rodents and other animals that have been killed or have died inside the attics, wall voids, or chimneys of structures. Some species are strong fliers, and all are attracted to bright light. The major health concerns associated with these flies is their ability to transmit diseases and cause human and animal myiasis, an infestation of living or dead tissue by fly larvae.

Control: The initial inspection should focus on identification of the fly (adult and/or larva) causing the problem and location of all resting and larval development sites. Because the adults often rest in breeding areas, it is helpful to inspect at night. Sanitation, or source reduction, is the most important step in blowfly control because it eliminates larval breeding sites. When successful, it significantly reduces the need for pesticide applications. If trash receptacles are the problem, property owners should be instructed to empty and clean them at least weekly to disrupt the developmental cycle.

Mechanical control measures include insect-proof garbage containers, self-closing doors, screening, caulking and air curtains. Electric fly traps, sticky traps and other devices should be used to reduce adult fly populations indoors and out. Insecticide applications should be directed at adults because sanitation and removal are the best control measures for larval breeding sites. Baits, aerosols and residual insecticides are the most widely used technologies used for fly control. The most widely used pesticides are emulsifiable concentrates, wettable powders and microencapsulated formulations. Most residual applications are made to the daytime and evening resting sites. Ultra-low volume and aerosol applications should be made when the adults are most active and there is the least risk of drift, surface contamination, and human or animal exposure.

Blow/Bottle Fly

CLUSTER FLY

Order/Family:
Diptera/Calliphoridae

Scientific Name:
Pollenia rudis (Fabricius)

Description: Cluster flies are close relatives of blow flies and are similar in size to house flies (3/8-inch) but are more robust in body structure. They are nonmetallic gray, lack stripes on the thorax (segments with the wings and legs attached), and have yellow or golden hairs on the back, behind the head, and around the base of the wings. Cluster flies appear narrow when at rest because their wings completely overlap over their backs. The larvae are typical spindle-shaped maggots but are never seen because they develop as parasites in earthworms.

Biology: The female flies mate in the spring and lay their eggs in soil crevices. The eggs hatch in three days and the larvae burrow into the bodies of earthworms where they develop. Development (egg to adult) requires 27 to 39 days. There are usually four generations per year.

Habits: Cluster flies are annoying because they overwinter as adults in the attics and wall voids of structures, especially older frame buildings. The common name of this species reflects its behavior of gathering in clusters before hibernation. They enter structures in early fall to seek shelter from cooling temperatures. Soon, a "cluster" of adult flies accumulates in wall voids and dark corners, under shelving, beneath curtains, and in other protected areas.

On warm days in winter and spring, they annoy building occupants when they become active and crawl sluggishly over walls or windows. When the weather warms, the cluster flies emerge from their hiding places and either exit the building or enter interior areas. They are stimulated by warmth and are often found on the south and west sides of buildings. Once stimulated, cluster flies are attracted to light.

Control: There is no effective control of the larval stage of these flies because they develop in earthworms. Control tactics for cluster flies should be initiated before they enter buildings in large numbers. The most effective long-term control in structures attractive to overwintering adults is to seal entry points in the walls and roof of the structure. To prevent entry into interior rooms, entry routes, e.g., around window pulleys, electrical outlets, switch boxes, and window and door frames should be sealed. Large accumulations of these flies can be removed with a vacuum cleaner.

During the winter, a bare light bulb in an attic will cause the flies to die from cold exposure and exhaustion of their food reserves. Insect light traps can also be placed in attics but require frequent servicing.

Cluster flies can not be controlled by disrupting the life cycle of the larvae (maggots) because they are parasitic on earthworms which are beneficial contributors to the environment. Residual applications of micro-encapsualted and wettable powder formulations should be applied in the fall to the exterior surfaces of structures in order to control these flies prior to entry. Applications should target southern and western exposures. Although flies in wall voids can be controlled using aerosol or dust formulations, it is not recommended because the dead flies may attract larder beetle or carpet beetle larvae and, thus, result in a worse problem.

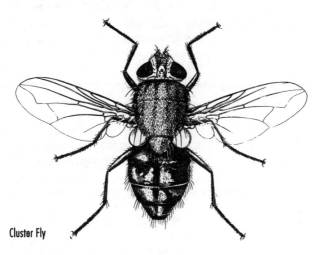

Cluster Fly

DRAIN/MOTH FLIES

Order/Family:
Diptera/Psychodidae

Scientific Name:
Psychoda spp. and *Telmatoscopus* spp.

Description: Moth flies are very small delicate and "hairy" flies about 1/16- to 1/4-inch long. They are yellowish, brownish-gray, or blackish. Their pointed, leaf-shaped, white-speckled wings are held roof-like over their backs when at rest. The body and wings are covered with long dense hairs. These flies have long antennae with 13 to 15 bead-like segments. Mature larvae are 1/8- to 3/8-inch long. They are aquatic and are long and cylindrical with the fore end of the body somewhat flattened on the lower side which has eight suckers. They have dark, hardened patches on the back of each segment, and they breathe through a hardened, stalk-like siphon tube at the end of the body.

Biology: The females lay 30 to 100 eggs in the jelly-like film that covers the stones in sewage treatment plant trickling filters or that line the water-free portions of drain pipes. The eggs hatch in two days, and the larvae complete development in nine to 15 days. The larvae feed on algae, fungi, bacteria, and sludge. The pupal stage lasts about a day and a half. The developmental time (egg to adult) is seven to 28 days. Adults live for up to two weeks.

Habits: Moth flies or drain flies become an annoyance within some structures when they breed in the liquids found in drains, dirty garbage containers, and septic tanks. They are a problem when they breed in large numbers in the filter beds of sewage treatment plants. Adult flies are poor fliers and are found in great numbers on walls or flying weakly in the area where they developed. Adults are more active at night and are seen hovering near the breeding site. During the day, they rest on vertical surfaces indoors and in protected areas outside. Adults feed on nectar and polluted water.

Control: When moth flies or drain flies are a persistent problem within a structure, it is evident that they are either developing in the building or at a nearby source such as a sewage treatment plant. Whenever practical, the source of infestation should be found and eliminated mechanically. Infestations which develop in drains can be eliminated by scrubbing the area with a brush and sink-cleaning materials followed by very hot water.

It is important to scrub the drain all the way to the trap. If the trap is not holding water, it should be repaired. To prevent water evaporation in the trap of seldom-used drains, a small amount of vegetable or mineral oil should be placed on the surface of the trap water. Large numbers of adult flies are easily controlled using an aerosol spray.

Large populations of flies which are developing in sewage filter beds can be controlled with periodic flooding of the bed which eliminates some of the larvae. Weed control in this area reduces the availability of resting sites. To control adult flies at outdoor resting sites, treat the area with a wettable powder or microencapsulated insecticide.

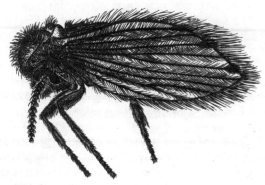

Drain/Moth Fly

FRUIT FLIES

Order/Family:
Diptera/Drosophilidae

Scientific Name:
Drosophila spp.

Description: Adult fruit flies are 1/8-inch long and dull yellow-brown to dark brown. Some species have distinctive red eyes, and the wings have two "breaks" in the leading edge (vein) nearest the body. The larvae are small (1/10- to 1/5-inch long) and very distinctive with an extended, stalk-like breathing tube at the rear of the body. The pupae are brown and seed-like with two horn-like stalks at one end.

Biology: The eggs are laid onto the surface of fermenting fruit or vegetables, or areas where moisture and yeast are abundant. The eggs hatch within 30 hours. Each female produces up to 500 eggs. The larvae complete development in five to six days and crawl to drier areas of the food or elsewhere in order to pupate. The life cycle (adult to adult) requires eight to ten days.

Habits: Fruit flies are common structural pests frequently associated with fermenting fruits and vegetables. They easily develop in over-ripe fruits or other foods, fermenting liquid in the bottom of garbage cans, a dirty mop, or a rotting potato or onion in the vegetable bin. Recycling bins and their contents and fruit and salad bars are ideal habitats and have resulted in increased problems with this pest fly. Recently-emerged adults are attracted to light.

Control: Fruit flies are best controlled by finding and eliminating the breeding material. The presence of adult flies indicates that the larvae are developing in some nearby fermenting materials. Complete and thorough sanitation is necessary to eliminate the source of the infestation. Insect light traps and baited jar traps fitted with tops which permit fly entry and prevent escape are effective in reducing the population but are no substitute for sanitation. Several species are small enough to pass through typical screens, so a smaller mesh maybe required. In limited situations, an aerosol can be used to knock down adults.

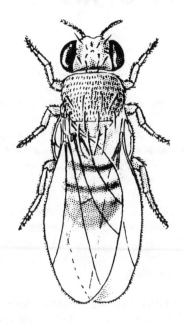

Fruit Fly

FUNGUS GNATS

Order/Family:
Diptera/Sciaridae and Mycetophilidae

Scientific Name:
Various

Description: Fungus gnats are very small flies 1/32- to 7/16-inch long, slender to robust, long-legged and mosquito-like. They are usually black, brown or yellowish. The dark winged fungus gnat (*Sciaridae*) has smoke colored wings, and the other fungus gnats (*Mycetophilidae*) have spots on their wings. The larvae are inconspicuous although they have well developed heads, 11 to 12 body segments, and a ventral bottom lobe on their last abdominal segment.

Biology: Little is known about the biology of either family. The dark winged fungus gnats molt four times as a larva. The other fungus gnats lay their eggs directly on larval food. The eggs hatch in a few days, and the larvae molt five times in six to eight days. They pupate in the ground and the adults emerge in three days.

Habits: The larval food consists of moist decaying organic matter and fungi growing in the soil. These gnats breed in various settings, such as rotting wood, animal waste, under bark, in over-watered plants and their saucers, and atriums. Flat-roof structures and accumulated droppings in bird cages are potential breeding sites.

Accumulations of old mulch around buildings provide an ideal habitat. These gnats are small enough to pass through typical structural screening. Adults are more active at dusk and are usually found near breeding sites. During the day, they seek out dark moist places to rest. They are attracted to light and often are seen at windows.

Control: Because there are hundreds of small fly species, the fly must first be identified before any course of action can be taken. The key to eliminating a gnat problem is location and removal of the breeding site. During the inspection, look for any site or condition which could support fungal growth. If adults are found around windows and doors, look outside for potential breeding sites, and remember that adults may be attracted to exterior lights.

Once the breeding site is located, the affected materials should be removed or dried out. Interior plants should not be overwatered but allowed to droop before watering; saucers and pots should be cleaned on a regular basis. Insect light traps can be used to significantly reduce indoor populations of these gnats. After the breeding site is eliminated, an aerosol application should be used to kill the remaining adults.

Fungus Gnat

HOUSE FLY

Order/Family:
Diptera/Muscidae

Scientific Name:
Musca domestica Linnaeus

Description: House flies are 1/8- to 1/4-inch long. They are dull gray with four dark stripes on the back of the thorax (segments right behind the head with legs and wings attached). They have two wings; the fourth longitudinal wing vein has a sharp upward turn. The head is dominated by large red-brown compound eyes which are surrounded by a light gold stripe. Short antennae emerge from between the eyes. They have sponging mouthparts. Mature house fly larvae or maggots are spindle shaped and creamy white. They have dark mouth hooks at the head end and breathing slits that look like a "wavy W" at the larger round tail end. House fly larvae are 1/4- to 3/8-inch long when they change to the brown seed-like pupal stage.

Biology: Female house flies lay their eggs singly but in clusters of 75 to 150 eggs in a variety of moist, rotting, fermenting, organic matter including animal manure, accumulated grass clippings, garbage, spilled animal feeds, and soil contaminated with any of the above items. A female may lay more than 500 eggs in a lifetime. The eggs hatch within a day, and the young larvae burrow into the breeding medium and complete development in three days to several weeks depending on the temperature and quality of food materials. Larvae migrate to drier portions of the breeding medium to pupate for three days to four weeks before emerging as adults. Under optimum conditions, house flies can complete their entire life cycles in less than seven days.

Habits: The adult flies may migrate to uninfested areas up to 20 miles away, but most stay within one or two miles of the breeding site. Adult house flies have a general appetite, feeding on foods ranging from excrement to human food. They feed on liquids but can eat some solid foods by liquefying it with regurgitated digestive tract fluids. During the day, house flies rest less than five feet above the ground and at night they rest above this height. House flies have been associated with many filth-related diseases, and, thus, are a significant health concern.

Control: The initial inspection should focus on identification of the fly (adult and/or larva) causing the problem and location of all resting and larval development sites. Because the adults often rest in breeding areas, it is helpful to inspect at night. Sanitation, or source reduction, is the most important step in house fly control because it eliminates larval breeding sites. When successful, it significantly reduces the need for pesticide applications. If trash cans are the problem, property owners should be instructed to empty and clean them at least weekly.

Mechanical control measures include insect-proof garbage containers, self-closing doors, screening, caulking and air curtains. Operating insect light traps indoors and at night is effective in controlling adult flies inside the structure. Sticky traps and other devices are also available to reduce adult fly populations indoors and out.

Any insecticide applications should be directed at adults because sanitation and removal are the best control measures for larval breeding sites. Products used for fly control include baits, aerosols and residual insecticides. The most widely used pesticides are emulsifiable concentrates, wettable powders and microencapsulated formulations. Most residual applications are made to the daytime and evening resting sites. Ultra-low volume and aerosol applications should be made when the adults are most active and there is the least risk of drift, surface contamination, and human or animal exposure.

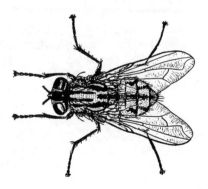

House Fly

BOXELDER BUG

Order/Family:
Heteroptera/Rhopalidae

Scientific Name:
Boisea trivittata (Say)

Description: Adult boxelder bugs are 1/2-inch long and brown-black with three red stripes on the thorax (segments with the legs attached) and red veins in the wings. The nymphs are smaller and are bright red. Boxelder bugs are found throughout the United States east of Nevada.

Biology: The adults overwinter in dry, protected locations, emerging in the spring to lay small, red eggs in the cracks and crevices in the bark of boxelder trees. The nymphs hatch in approximately two weeks when new leaves appear. The young bugs suck the juice out of the tree leaves and twigs with their piercing-sucking mouthparts. They molt five times before becoming adults. In warm climates, these bugs have two generations per year.

Habits: They prefer to feed on the leaves, twigs, and seeds of female boxelder trees and also on maple, ash, and the young fruit of grapes, apples, and plums.

Boxelder bugs do little apparent damage to the boxelder tree. They become a nuisance around structures when they attempt to enter to find overwintering sites. Their migration begins in the autumn when they congregate on the south side of structures, rocks, and trees in areas warmed by the sun. Subsequently, they may fly to an adjacent building, enter it, and hibernate for the winter.

Indoors, their droppings stain drapes, curtains, furniture, sheers, and other materials where they rest. If handled, boxelder bugs can bite, and when crushed, they emit a strong disagreeable odor.

Control: The most effective control of boxelder bugs is to prevent their entry into structures. Removal of female boxelder trees usually eliminates the problem but is rarely practical if trees are left on adjoining properties. If removal is impractical, the trees should be treated with a contact or systemic insecticide designed to kill the bugs while they are still nymphs feeding on the tree.

Entry into buildings should be prevented by sealing and caulking gaps around siding, windows, doors, pipes, wires, etc. To be effective, this work should be completed by early August. Indoors, vacuuming removes accessible boxelder bugs. The vacuum bag should be taped and discarded.

Microencapsulated and wettable powder products are the most effective formulations for outdoor application sites. These products should be applied as a three- to ten-foot band around the perimeter of the structure and/or around potential entry points. Timing of this application is critical, e.g., in the Northeast, mid- to late August is the ideal treatment time.

Treating wall voids to control boxelder bugs is not recommended because the dead bugs may attract secondary pests, such as dermestid beetles, thus, making a bad problem worse. It is preferable to wait until spring when the adult bugs emerge and then all points of entry should be sealed.

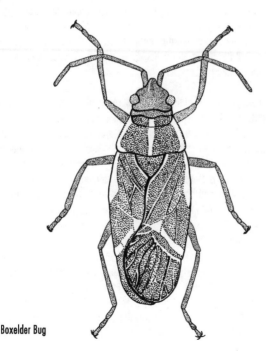

Boxelder Bug

CENTIPEDES

Class/Order:
Chilopoda/Various

Scientific Name:
Various

Description: Adult centipedes are yellowish to dark brown, often with dark markings, and 1/8- to 6-inches long. The body is flattened with 15 to 177 body segments which typically have one pair of legs each. They have one pair of slender antennae.

The house centipede is grey-yellow with three stripes down the back and has very long legs banded with white. The largest centipedes are found in the Southwest.

Biology: Centipedes typically overwinter outdoors, and, in the summer, lay 35 eggs or more in or on the soil. Newly hatched centipedes have four pairs of legs; during subsequent molts, the centipede progressively increases the number of legs until becoming an adult. Adults of many species live a year and some as long as five to six years.

Habits: Centipedes, including the house centipede, prefer to live in moist environments. The house centipede can live indoors in damp basements, moist closets, and bathrooms and outdoors under stones, decaying firewood, objects on the ground, piles of leaves, mulch, etc. Most centipedes are active at night.

The first pair of legs on centipedes has poison glands which are used to kill prey, such as insects and spiders. They obtain most of their water from their prey. Centipedes can bite humans, but the bite is seldom worse than a bee sting.

Control: Centipedes which are a nuisance outdoors can be controlled by removing harborage areas such as piles of trash, stones, boards, leaves, grass, and compost. Entry into buildings should be prevented by sealing and caulking gaps around siding, windows, doors, pipes, wires, etc.

Microencapsulated and wettable powder products are the most effective formulations in the moist habitats preferred by centipedes. These products should be applied as a three- to ten-foot band around the perimeter of the structure, into harborage sites, and/or around potential entry points.

House centipedes can be controlled indoors by eliminating their harborage areas where possible. A vacuum should be used to remove exposed centipedes. Liquid and dust applications can be introduced into wall voids, cracks and crevices along baseboards, and into other potential hiding places.

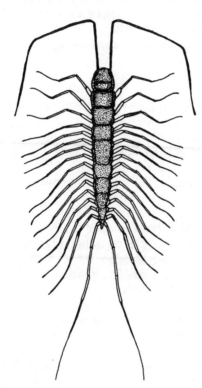

Centipede

CLOVER MITE

Order/Family:
Acari/Tetranychidae

Scientific Name:
Bryobia praetiosa Koch

Description: Clover mite adults are about 1/64-inch long and brown to olive-green in color. Their body shape is similar to that of ticks. The adults and nymphs have eight legs. They are easily distinguished from other mites by their very long front legs, which are longer than the body and twice as long as any of the other legs. The younger stages are bright red, as are the eggs, which are smooth and round.

Biology: Females are parthenogenetic, laying eggs without fertilization by a male. Approximately 70 eggs are deposited in the fall in protected locations on building foundations and under the bark of trees. The eggs do not hatch unless the temperature is between 40° and 70°F. In the spring, they hatch into the six-legged larval stage which then molts into the protonymph followed by the deutonymph, which, in turn molts into the adult stage. Developmental time (egg to egg) takes from one to seven months depending on environmental conditions.

Habits: Clover mites are plant feeders that have been found infesting more than 200 different plants. They can overwinter as adults, immatures, or eggs. They build up very large populations around structures surrounded with lush, well-fertilized lawns and shrubbery. They often move into buildings in tremendous numbers in the autumn when vegetation begins to die.

In the spring, large numbers indoors usually is the result of recent mulching and onset of higher temperatures. Large populations of clover mites occur on the flat roofs of commercial buildings and are associated with moss growth. Clover mites are harmless but are a great annoyance to building occupants. If crushed, they leave a red stain on walls, floors, or furnishings.

Control: Infestations in structures can be controlled using mechanical control measures. Large populations of mites should be removed using a vacuum cleaner. Residual or aerosol pesticide applications indoors does little to solve

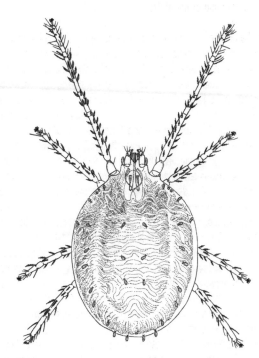

the problem and are not recommended.

Clover mite invasions can be prevented by eliminating lush vegetation in a 18- to 24-inch band around the perimeter of the building. A micro-encapsualted or wettable powder formulation should be applied approximately two feet up the foundation, or to where the siding begins, and to the soil surface in a three- to ten-foot band around the structure. All exterior cracks and other potential points of entry should be treated. Shrubbery and tree trunks within this area should also be treated. To be successful, product applications should be made in early May and again in late August.

Clover Mite

EARWIGS

Order/Family:
Dermaptera/Various

Scientific Name:
Various

Description: Adult earwigs are 1/4- to 1-inch long, dark brown to black, with a red head and pale yellow-brown legs. The body is long and flattened. Earwigs usually have two pairs of wings, the hind wings being fully developed and folded beneath the short, leathery front wings. The thread-like antennae are half as long as the body. The most notable characteristic are pincer-like appendages at the end of the abdomen, the forceps.

Biology: The female lays several batches of approximately 50 eggs in a nest-like shallow depression beneath a board or stone. Those laid in the winter hatch in about 72 days; those laid in the spring hatch in 20 days. The nymphs look much like the adults and molt four to five times before becoming adults, which takes about 56 days.

Habits: Earwig females are interesting because they display a mothering instinct, protecting the nest and the nymphs until they have reached their second molt. Earwigs usually live outdoors and feed on plant material. They are very general feeders and seldom do a great deal of damage to any particular plant. They are active at night, hiding during the day under stones and other objects.

Earwigs are outdoor insects which become household pests when they invade structures, usually in the fall or at night. Indoors, they are usually found in cracks and crevices and under furniture and carpeting. They are considered pests because of their presence and because they have a foul odor when crushed. Some species of earwigs are attracted to lights.

Control: Earwig control begins outdoors by removing moist harborages, such as wood piles, landscape timbers, stones, rocks, etc. The yard should be mowed and weeded, and flower beds should not be over mulched. It is very helpful if an 8- to 24-inch vegetation free zone is left adjacent to the foundation.

Entry into buildings should be prevented by sealing and caulking gaps around siding, windows, doors, pipes, wires, etc. Yellow bug-lights or sodium vapor lighting is less attractive to earwigs. A vacuum should be used to remove accessible earwigs.

Baits are effective when applied indoors as a band around structures and/or directly into harborage areas. Microencapsulated and wettable powder products are the most effective formulations in the moist habitats preferred by earwigs. These products should be applied as a three- to ten-foot band around the perimeter of the structure, into harborage sites, and/or around potential entry points.

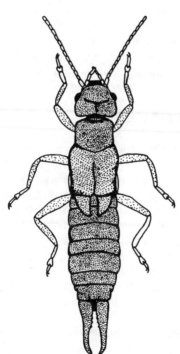

Earwig

ELM LEAF BEETLE

Order/Family:
Coleoptera/Chrysomelidae

Scientific Name:
Pyrrhalta luteola (Muller)

Description: Adult elm leaf beetles are about 3/16- to 1/4-inch long and yellow to olive green with a dark stripe down each side of their wing covers. There are usually four dark spots on the pronotum (segment right behind the head). The eggs are orange, and spindle-shaped, and the larvae are worm-like, black or black and yellow, and up to 1/2-inch long.

Biology: Adult elm leaf beetles overwinter in protected locations, often in houses or other structures. They emerge in the spring and move to elm trees where they lay their eggs in groups of five to 40 on the underside of leaves. Elm leaf beetle larvae often move to the base of the trees in large numbers to pupate. Developmental time (egg to adult) is approximately 38 days. There are two to five generations per year.

Habits: The adults and larvae feed on elm leaves, skeletonizing them and cuasing the leaves to appear net-like. The droppings from this feeding activity can damage car paint on hot summer days. Elm leaf beetle adults move into buildings in the fall to seek hibernation sites. When the adults move indoors in large numbers, they can be found under siding, in attics, behind drapes, curtains, and paintings, between books, and in other protected locations. Stains may occur if they are crushed.

Control: The most effective way to eliminate elm leaf beetles is to control the larvae and adults while they are still on the tree. Trees should be treated with a contact, residual, or systemic insecticide designed to kill the young beetle larvae while they are feeding on the tree. Another effective strategy is to treat three to four feet up the base of the trees to kill those larvae migrating to pupate.

The most effective control of elm leaf beetles is to prevent their entry into structures by sealing and caulking gaps around siding, windows, doors, pipes, wires, etc. To be effective, this work should be completed by early August.

Indoors, a vacuum can be used to remove accessible elm leaf beetles. The vacuum bag should be taped and discarded.

Microencapsulated and wettable powder products are the most effective formulations for outdoor application sites. These products should be applied as a three- to ten-foot band around the perimeter of the structure and/or around potential entry points. Timing of this application is critical, e.g., in the Northeast, mid- to late August is the ideal treatment time.

Treating wall voids to control elm leaf beetles is not recommended because the dead beetles may attract secondary pests, such as dermestid beetles, thus, making a bad problem worse. It is preferable to wait until spring when the adult beetles emerge, then seal all points of entry.

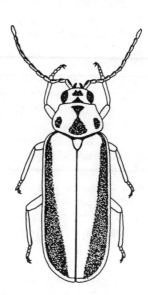

Elm Leaf Beetle

FIELD CRICKETS

Order/Family:
Orthoptena/Gryllidae

Scientific Name:
Gryllus spp.

Description: There are more than 25 different species of field crickets. These insects range from 1/2- to 1-1/8-inch long and have the typical stout body with large "jumping" hind legs characteristic of crickets. Field crickets are usually black but can also be brown or straw-colored. They have slender antennae which are much longer than the body. The wings on adult crickets lay flat on the back and are bent down on the sides. Adult female crickets have a long slender, tube-like structure (ovipositor) projecting from their abdomen which they use to lay eggs. Both males and females have two antenna-like structures (called cerci) that are attached to the sides of the tip of the abdomen. Nymphs look like the adults but are smaller and the wings are not fully developed.

Biology: Most field crickets overwinter as eggs that have been laid in moist, firm soil or occasionally as nymphs. Each female lays 150-400 eggs which hatch in the spring. The nymphs molt eight or nine times, usually becoming adults in about 78-90 days. There maybe several generations per year.

Habits: Field crickets can be important agricultural pests. They can also become household problems in late summer when they move out of fields and into buildings. Field crickets can damage furniture, rugs, and clothing, and the chirping of the adult males is irritating to some individuals. These insects are unable to survive indoors for long periods and usually die off by winter.

Field crickets are active at night, hiding in dark warm places during the day. They are attracted to lights, often by the thousands.

Control: Field cricket control begins outdoors by removing moist harborages, such as wood piles, landscape timbers, stones, rocks, etc. The yard should be mowed and weeded, and flower beds should not be overmulched.

Entry into buildings should be prevented by sealing and caulking gaps around siding, windows, doors, pipes, wires, etc. Yellow bug-lights or sodium vapor lighting should be used outside to avoid attracting crickets to doors or windows. A vacuum can be used to remove accessible crickets.

Baits are effective when applied indoors, as a band around structures and/or directly into harborage areas. Microencapsulated and wettable powder products are the most effective formulations in the moist habitats preferred by crickets. These products should be applied as a three- to ten-foot band around the perimeter of the structure, into cricket harborage sites, and/or around potential entry points.

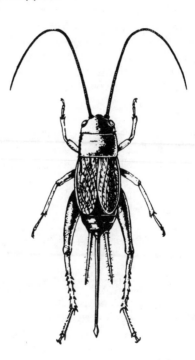

Field Cricket

HOUSE CRICKET

Order/Family:
Orthoptera/Gryllidae

Scientific Name:
Acheta domesticus (Linnaeus)

Description: House crickets are about 3/4- to 7/8-inch long and yellow-brown or straw-colored with three dark bands across the top of the head. The house cricket has long, slender antennae that are much longer than the body. The wings on adult crickets lay flat on the back and are bent down on the sides. Adult female crickets have a long slender, tube-like structure (ovipositor) projecting from their abdomen which they use to lay eggs. Both males and females have two antenna-like structures (called cerci) that are attached to the sides of the tip of the abdomen. Nymphs look like the adults but are smaller with less developed wings.

Biology: Outdoors, female house crickets lay an average of 728 eggs in protected areas. Eggs are the overwintering stage outdoors. The eggs hatch in late spring, and nymphs reach the adult stage by late summer. House crickets in the wild have only one generation per year.

Indoors, house crickets lay approximately 104 eggs in cracks, crevices, and other dark areas including behind baseboards. The nymphs undergo seven to eight molts and complete development in 53-56 days. House crickets can complete their life cycle indoors and live indefinitely in homes or other structures.

Habits: House crickets are seldom a major problem in structures as they prefer to live outside during warm weather. They move indoors when it gets colder and to find moisture. House crickets damage clothing and other fabrics including synthetics. They eat large holes in fabric as opposed to smaller holes caused by common fabric pests. Some people object to their presence and the chirping noise produced by males as they rub their wings together.

Crickets may be introduced unintentionally into structures as they are brought in as food for pet snakes or other animals and escape. Crickets are active at night, hiding in dark warm places during the day. They are attracted to lights, often by the thousands.

Control: House cricket control begins outdoors by removing moist harborages, such as wood piles, landscape timbers, stones, rocks, etc. The yard should be mowed and weeded, and flower beds should not be over mulched.

Entry into buildings should be prevented by sealing and caulking gaps around siding, windows, doors, pipes, wires, etc. Yellow bug-lights or sodium vapor lighting should be used outside to avoid attracting crickets to doors or windows. A vacuum should be used to remove accessible crickets.

Baits are effective when applied indoors, as a band around structures and/or directly into harborage areas. Microencapsulated and wettable powder products are the most effective formulations in the moist habitats preferred by crickets. These products should be applied as a three- to ten-foot band around the perimeter of the structure, into cricket harborage sites, and/or around potential entry points.

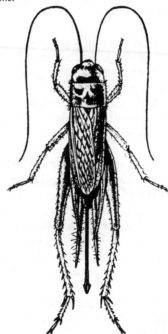

House Cricket

HOUSE DUST MITES

Order/Family:
Acari/Pyroglyphidae

Scientific Name:
Dermatophagoides spp.

Description: House dust mites are very small creatures about 1/64-inch long. Their presence is often suspected before they are actually seen and identified. The two most common species are *Dermatophagoides farinae,* the North American house dust mite, and *D. pteronyssinus,* the European house dust mite.

Adult mites are clear, and their cuticle has simple striations that can be seen from both the top view and from the bottom view. The bottom view of the mite also reveals long hairs extending from the outer margins of the body and shorter hairs on the rest of the body.

Biology: Adult females lay 66 to 68 eggs singly or in small groups. After egg hatch, a six-legged larva emerges. After the first molt, an eight-legged nymph appears. After two nymphal stages occur, a eight-legged adult emerges. The life cycle from egg to adult is about one month. The adult lives for an additional one to three months. Their primary food source is dander (skin scales) from humans and animals, but pet food, fungi, cereals, food stains and crumbs also provide needed nutrients.

Habits: House dust mites inhabit buildings with micro-habitats that sustain a relative normal to high moisture content. Occupied rooms that are dry (below 40% relative humidity) and well ventilated probably would not support this mite, whereas a damp room or its adjoining room will support a mite population. Even in relatively dry rooms, the bottom inch of air above the floor is cooler and slightly more humid. The mites, their body parts, and excretions are suspected of being allergens to over 500 million people worldwide including 50-80% of asthmatics.

Control: The house dust mite is virtually invisible to the unaided eye. Inspection for and identification of these tiny arthropods and their body fragments requires a microscope to search through dust debris collected on cushions and mattresses, in pet bedding, cracks and crevices, and other areas where dander can collect. Service for house dust mite control often is requested by clients who have been diagnosed by physicians as allergic to the house dust mite and/or the allergens it produces.

Mite populations can be eliminated by reducing the humidity below 45% and improving ventilation. However, the allergenic materials persist even if the population is destroyed. Allergy reduction is dependent on removing these particles. It is extremely difficult to remove every pocket of dust or to stop its daily accumulation. Unless equipped with a HEPA filter, vacuum cleaners blow microscopic mite fragments into the air, often aggravating the allergy.

Nevertheless, effective dust control measures do exist. Suggest options such as encasing mattresses and pillows with plastic covers, replacing feathered pillows with synthetic ones, removing stuffed animals, and installing air cleaners and filters. Some furnace and air conditioner filters trap dust with an electrostatic charge. An allergist may recommend that customers reduce dust accumulation by replacing carpeting with wood, tile, or linoleum flooring.

Currently, few, if any, pesticides are labeled for house dust mites. Several non-pesticide products are available for treatment of house dust mites and their allergenic components; however, their effectiveness is questionable. The active ingredient of each is benzyl benzoate and tannic acid, respectively.

House Dust Mite

LADY BUGS

Order/Family:
Coleoptera/Coccinellidae

Scientific Name:
Various

Description: Ladybugs (beetles) are hemispherical in shape and are 1/16- to 1/4-long. They are brightly colored. Most are red, brown, or tan with black spots, while a few are black with red spots. The adults have three tarsal segments (part of the leg farthest from the body). This characteristic distinguishes them from destructive beetles of similar size and shape which have four tarsal segments.

Biology: Eggs are orange and are laid on end in single or multiple groups of 12 on plants infested with aphids. The larvae are flattened and spindle-shaped, with warts or spines on the dorsal (back) side. They undergo four molts and are brightly colored with blue, black and orange. The pupae, which do not form cocoons, are attached to leaves by the tip of their abdomens.

Habits: These insects are beneficial because the larvae and adults eat a variety of outdoor ornamental pests, such as aphids, mealybugs, whiteflies, scale, other soft bodied insects and their eggs.

In the fall, the adults seek protected areas to overwinter, preferring areas beneath rocks, bark, leaves and landscape timbers; however, occupied structures are also suitable. Adults are attracted to light and are often seen in window sills and light fixtures.

Control: Because ladybugs are beneficial insects and pose no threat to health or property, no direct control methods are recommended. Educating the customer and preventing entry into structures by sealng up all external cracks, gaps in siding, and openings around door and window frames are the best solutions. Similar areas on the inside of the structure should be sealed to prevent them from entering the interior from wall voids and voids between floors. Screens should fit tightly and unscreened doors and windows should not be left open. Soffit and ridge vents which are also points of entry should also be screened.

Once the beetles are inside, the best course of action is to use a vacuum cleaner for removal, and then remove, tape and discard the vacuum bag, or release the beetles outside away from the structure. Light traps can be used effectively in some situations to reduce indoor infestations. Spraying indoors for the beetles is of no value and provides little relief from the problem.

Periodic inspection of exterior plants for aphids, scale, and other soft-bodied insects helps to anticipate the problem. Preferably, these plants should be treated with a systemic insecticide or a topical formulation to control ladybug prey and minimize their interest around the structure. Avoid spraying if any of the beetle life stages are present. The beetles are often as effective in controlling the ornamental pests as are insecticides.

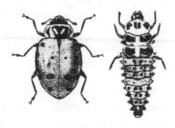

Lady Bug

MILLIPEDES

Class/Order:	Scientific Name:
Diplopoda/Various	Various

Description: Most millipedes are brown or black, but some species are orange or red. They range in size from 5/8-inch to 4-inches in length. There are many species of millipedes, most of which are long, cylindrical, many-segmented, worm-like creatures. However, a few millipedes appear to be flattened. Every millipede has two pairs of legs attached to each apparent body segment.

Biology: Females lay 20 to 300 eggs in nests in the soil. The eggs hatch after several weeks. The young initially have only three pairs of legs and seven segments. They undergo a series of seven to ten molts during which the number of legs and segments increase. Many species reach sexual maturity in two years, but some require four or five years to complete development. Adults can live several years. Most species overwinter as adults or young.

Habits: Millipedes are found outdoors in situations where there is moisture and decaying organic matter, such as under trash, grass clippings, mulch, rotting firewood, leaf litter, etc. They can build up tremendous populations in forest litter and compost heaps. Millipedes become structural pests when they invade homes and other structures, sometimes in staggering numbers. This occurs when standing water in their natural habitat forces them out. Indoor movement is also caused by drought and their natural migratory and mating instincts in the fall. Millipedes usually die within a few days of entering a structure unless there is a source of high moisture and a food supply.

Millipedes are active at night. They scavenge feeding on decaying organic matter. Most species produce a foul smelling fluid that comes out the sides of their bodies and is toxic to some insects and small animals. This substance can cause blisters on human skin.

Control: Millipede control begins outdoors by removing moist harborages, such as wood debris, rocks, grass clippings, and leaf litter, decaying and other accumulations of organic material, where possible. Firewood should be stored off the ground. The lawn should be dethatched, mowed, and edged to promote drying. Lawns should be watered in the morning to promote drying by the afternoon. Flower beds should not be over mulched.

Entry into buildings should be prevented by sealing and caulking gaps around siding, windows, doors, pipes, wires, etc. Large numbers of these structure-invading pests are easily controlled by vacuuming and discarding the collected material.

Microencapsulated and wettable powder products are the most effective formulations in the moist habitats preferred by millipedes. These products should be applied as a three- to ten-foot band around the perimeter of the structure, into harborage sites, and/or around potential entry points. Infested grassy areas should be treated to reduce outdoor populations. When large numbers of millipedes invade structures residual applications into harborage might be necessary. Dusts can be used in dry crawlspaces.

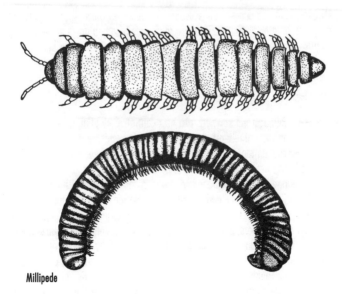

Millipede

PILLBUGS / SOWBUGS

Class/Order:
Crustacea/Isopoda

Scientific Names:
Armadillidum vulgare (Latreille) — Common Pillbug
Porcellio laevis Koch — Dooryard Sowbug
Porcellio scaber (Latreille)

Description: Pillbugs and sowbugs are dark gray, 1/4- to 5/8-inch long, oval crustaceans that are humpbacked, appear to be covered with segmented armor, and have seven pairs of similar legs. Pillbugs can be distinguished from sowbugs because pillbugs are able to roll up into a ball when alarmed and lack the two prominent tail-like appendages that characterize sowbugs.

Biology: Female pillbugs and sowbugs carry their eggs in a "brood pouch" or marsupium where the young hatch in about 45 days. There are 24-28 eggs in each brood and one to three broods produced each year. The young molt every one to two weeks until they become adults. Adults live about two years.

Habits: Pillbugs and sowbugs usually are found in moist situations where they feed on organic matter. They are commonly found outdoors under stones, boards, and piles of plant material. Pillbugs and sowbugs invade basements, crawlspaces and sometimes other parts of structures that have higher-than-normal moisture. They cause no damage but are considered a nuisance. Unless disturbed, they are usually active only at night.

Control: Pillbugs and sowbugs are best controlled by eliminating the moist environment that initially attracts them. Piles of organic matter, dense ground cover near foundations, or ground level windows, boards, stones, flower pots, firewood, and other materials resting on the ground, serve as food sources, and harborage areas for pillbugs and sowbugs and they should be removed or modified to reduce the pillbug or sowbug population.

Entry into buildings should be prevented by sealing and caulking gaps around siding, windows, doors, pipes, wires, etc. Large numbers of pillbugs and sowbugs which successfully invade homes are easily controlled by vacuuming them up and discarding the collected material. Unfinished basements and crawlspaces should be well-ventilated to reduce moisture which is attractive to these pests.

Microencapsulated and wettable powder products are the most effective formulations in the moist habitats preferred by sowbugs and pillbugs. These products should be applied around the perimeter of the structure, into harborage sites, and around potential entry points. Application of these products indoors may be necessary and can be supplemented by the use of dusts in drier areas.

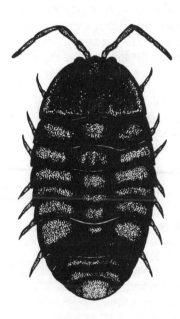

Pillbug

PSOCIDS/BOOKLICE

Order/Family:
Psocoptera/Various

Scientific Name:
Various

Description: Psocids, commonly called book lice or paper lice, are not lice at all. In fact, they look more like termite workers but are much smaller. They are 1/32- to 1/4-inch long, pale, and the wings, if present, are usually held in a roof-like position over the body when resting. The antennae are long and thread-like.

Biology: Psocids undergo gradual metamorphosis, i.e., the immature forms resemble the adult forms. The developmental time (egg to adult) ranges from 24 to 110 days, though this varies greatly with species and environmental conditions. Low relative humidity hinders development or causes death due to desiccation.

Habits: Indoors, psocids seek out damp, warm, undisturbed areas, such as inside books on shelves, moldy foodstuffs in a pantry, and stacks of stored paper goods. They are found in old window and door casings where moisture accumulates and mold grows. In new structures they are attracted to the moisture in fresh plaster. In warehouses, the most common complaints about this pest are associated with wooden pallets and palletized materials. Outdoors, they seek out protected areas, e.g., under tree bark or leaves, mulch, bird and rodent nests, etc. High moisture is essential to their survival.

Psocids feed on various substances including molds, fungi, starches, and dead insects. Certain species feed on the starchy glue and paste that binds books and wallpaper to surfaces. Other species feed on bulk grains that are moldy.

Control: Inspection for booklice is difficult because their presence often is unnoticed by the customer; they are difficult to detect without the aid of a hand lens. If wooden pallets are the source of the problem, they should be fumigated or replaced with plastic ones.

The most successful control measure for psocids is to reduce the relative humidity in infested areas to less than 50% which dries out infested goods and eliminates mold, mildew, and other food sources. Ventilation increases moisture evaporation, thus, eliminating the humid conditions that support microscopic mold and fungi growth on packaging materials, grains, and other products. Equipment, flooring, and walls should be cleaned with a disinfectant to remove mold and fungal growth.

ULV applications kill exposed psocids. However, they often are in protected areas and are not affected by this type of treatment. Therefore, crack and crevice treatments are most effective in areas such as book shelves or storage racks where psocid activity is observed. Products which contain silica aerogel and other types of desiccants are effective. In extreme cases, particularly in sensitive accounts, such as food and drug manufacturing facilities, fumigation of the site may be necessary.

Psocid

SPRINGTAILS

Order/Family:
Collembola/Various

Scientific Name:
Various

Description: Springtails are very small, whitish-gray or light-colored insects, measuring 1/32- to 1/8-inch long. They have a bulbous "humpbacked" body, no wings, and a distinctive head with long antennae. Springtails get their name from a forked appendage attached to the end of the abdomen which can be bent under the body and, when released, helps the insect to "spring" forward, much like a flea.

Biology: Females lay eggs singly or in clusters in moist areas. The immature springtails undergo five to ten molts before they become adults. The adult continues to molt up to 50 times throughout its life with no increase in size after the fifteenth molt. Developmental time (egg to adult) requires two to three months, and, occasionally, as long as two years.

Habits: Springtails are always found in very moist situations. Outdoors, their populations can reach as many as 50,000 per cubic foot of soil. They are typically associated with leaf litter, mulch, firewood, landscape timbers, potted plants, railroad ties, etc. Nineteen different species of these insects have been found to invade homes and buildings, doing so when their living area becomes too dry and they need a moisture source.

Some are small enough to enter through window screens. They can be found in sinks and basins, floor drains, around sweating pipes, in moist basements or crawlspaces, on moldy furniture, and in the soil of potted plants. These insects feed on decaying organic matter, algae, and fungi. They are attracted to light.

One species has been associated with itching skin of people who work in areas where large numbers of springtails are found. Many times this dermatitis is mistakenly blamed on fleas because of the way that springtails hop about.

Control: The best control for springtails is to eliminate the source of moisture that sustains them. Outdoors, they can be controlled by removing harborage areas and storing items off the ground. Drying out the insect's surroundings eliminates the pest. Particles of food and concentrations of dust and lint that may serve as a food source should be removed. Entry into buildings should be prevented by sealing and caulking gaps around siding, windows, doors, pipes, wires, etc.

Pesticide applications may be necessary in areas where large numbers of insects have concentrated and modification of the environmental conditions is not practical. Microencapsulated and wettable powder products are the most effective formulations in the moist habitats preferred by springtails. These products should be applied around the perimeter of the structure, into harborage sites, and/or around potential entry points.

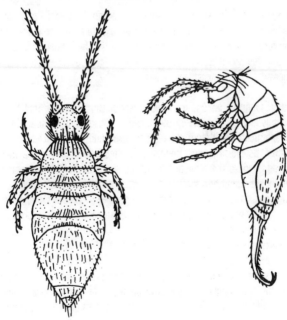

Springtail

CIGARETTE BEETLE

Order/Family:
Coleoptera/Anobiidae

Scientific Name:
Lasioderma serricorne (Fabricius)

Description: The cigarette beetle is light brown and 1/16- to 1/8-inch long. It has a humped appearance because its head and the first body segment is bent downward. It has saw-like antennae and smooth wing covers. The larva is a 1/16- to 1/8-inch long c-shaped white grub that has a somewhat fuzzy appearance because it is covered with long hairs.

Biology: The adult female lays 30-42 eggs which hatch in about six to 10 days in or near food material. The larvae feed on these foods and can complete development in five to 10 weeks. They pupate for two to three weeks in silken cocoons which are covered with food debris. The entire life cycle, egg to egg, requires 30 to 90 days. There are three to five generations per year, depending on the temperature. Adults usually live 23 to 28 days.

Habits: Cigarette beetles are major pests of stored tobacco, but they also commonly feed on all types of spices, books, upholstered furniture, dried fruit and vegetables, nuts, drugs, seeds, old rodent bait, and insecticides such as pyrethrum powder. The most frequently infested items in homes are dried dog food and paprika. These beetles readily penetrate packaging materials.

The adults are strong fliers and are more active in late afternoon and on cloudy days. They easily enter homes through open windows and doors and are frequently found along window sills. The presence of adults usually means the larvae are developing somewhere in the home.

Control: The first step in cigarette beetle control is to find all infested materials and remove or destroy these products. To prevent beetle entry, all screens should be repaired, exterior points of entry sealed, and exterior doors kept closed. Light traps can be used to monitor and control small populations of adult beetles.

Large numbers of adults are easily killed with aerosol applications. Liquid and dust formulations should be used to treat cracks, crevices and other voids in cabinets, shelves, pantries, and other locations where infestations are found. Applications should be made to insure that all potential harborage areas are well treated. These treatments are supplementary to the elimination of infested material and should never be used as the sole means of control. Cigarette beetles are controlled in tobacco storage facilities with sanitation, supplemented by frequent applications of aerosols and fumigation when necessary.

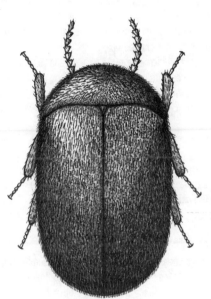

Cigarette Beetle

DRIED FRUIT BEETLE

Order/Family:
Coleoptera/Nitidulidae

Scientific Name:
Carpophilus hemipterus (L.)

Description: Adult dried-fruit beetles are about 1/8-inch long, oval in shape, and dull to shiny black. The tip of each wing cover has an easily-seen, amber-brown spot with a smaller, less easily-seen spot where the wings attach to the body. The wing covers are short and leave the last few segments of the abdomen exposed. These insects also have light red-brown to amber-colored legs and antennae with knob-shaped tips.

The larvae are 1/4-inch long and the head and tail are amber-brown. The larval body is covered with spine-like hairs and has two relatively large projections extending from the end of the abdomen with two smaller ones just in front of the large projections.

Biology: The female dried-fruit beetle lays an average of 1,000 eggs in her lifetime. Eggs are deposited on ripening fruit or piles of fermenting fruit, often while it is still on the tree or on the drying trays. Eggs hatch in one to seven days, and the active larvae develop in the fruit for five days to four months. They remain in the pupal stage for about six to 14 days. The life cycle, egg to adult, can be as short as 16 days to several months, depending on temperature.

Habits: Dried-fruit beetles are major pests of the dried-fruit industry. Both the larvae and the adults feed on fruit while it is being dried, preferring moist fruit over very dry products. These insects have been found to infest dried figs, plums, peaches, apricots, bananas, drugs, nuts, bread, biscuits, and grain. When found in packaged fruit, the presence of dead insects, molted skins, and masses of feces make the fruit inedible. The insects also introduce yeasts into drying fruit which cause it to sour.

Control: These beetles are seldom a problem in houses unless infested dried fruit is brought in from stores. The infestation is easily controlled by removing and destroying the infested material. No insecticide treatment is required other than the use of pyrethrins for the control of exposed adults.

Infestations in fruit undergoing drying or in stored dried fruit are best controlled with fumigations as necessary and intensive sanitation designed to prevent infestation in the first place.

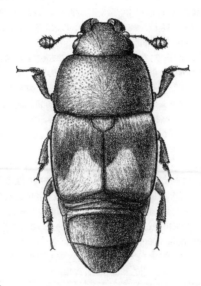

Dried Fruit Beetle

DRUGSTORE BEETLE

Order/Family:
Coleoptera/Anobiidae

Scientific Name:
Stegobium paniceum (Linnaeus)

Description: Drugstore beetles look almost identical to cigarette beetles. They are 1/16- to 1/8-inch long, light brown to red-brown, humpbacked in appearance, and the head is not visible. They differ from cigarette beetles in that they have pits on their wing covers arranged in long rows, whereas cigarette beetles have smooth wing covers. Drugstore beetle antennae end in a distinct club composed of three elongated segments, while cigarette beetle antennae are saw-like. Mature drugstore beetle larvae are 1/16- to 1/8-inch long, C-shaped, relatively hairless, and do not have the fuzzy appearance of cigarette beetle larvae.

Biology: Females lay their eggs singly in or near the food they are infesting. The eggs hatch within a few days. The larval period lasts four to five months, and the pupal stage lasts 12 to 18 days. The complete life cycle, egg to egg, requires about seven months. There may be four generations per year depending on the temperature. However, in houses, it is generally thought that drugstore beetles have one generation per year.

Habits: Drugstore beetles feed on all types of foods and spices, as well as on leather, wool, hair, books, drugs, and museum items. They readily penetrate packaging materials. The adults can fly and are attracted to light.

Control: Stored products and storage areas should be inspected carefully for evidence of infestation. Pheromone traps are available and can be used to identify areas of activity. Heavily infested materials should be discarded. If there is a light infestation, an alternative to disposal is heat treatment (140^0 to 176^0 F for several hours) which kills all life stages of the drugstore beetle.

Books and manuscripts, furniture, and other infested materials that are not easily replaced, might need to be fumigated in order to control drugstore beetles. To prevent reinfestation of food products by residual populations, liquid or dust formulations should be applied to surfaces, cracks, crevices, and other potential harborage areas. Aerosol applications can be used to temporarily reduce the exposed adult population. Product applications are supplementary to the clean-up of infested material and should never be used as the sole means of control.

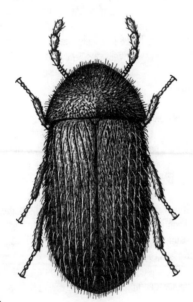

Drugstore Beetle

FURNITURE CARPET BEETLE

Order/Family:
Coleoptera/Dermestidae

Scientific Name:
Anthrenus flavipes LeConte

Description: Adult furniture carpet beetles are 1/16- to 1/8-inch long, oval in shape, and covered with spots of yellow, white, and black scales. The underside of the body is white.

The larvae are about 3/16-inch long, rather stubby-looking, and clothed with bands of stiff, erect brown hairs. The body is wide in the middle but tapers at both ends. Three thick tufts of hairs arise from either side of the body near the rear end.

Biology: Adult females lay about 30-100 eggs in areas where the larvae are able to feed. The eggs hatch in nine to 19 days, and the larvae develop from 70 to 94 days before pupating in their last larval skin. The adults emerge from the pupal skin in 9 to 19 days and may live for another 60 days. Development from egg to adult is 90 days to two years.

Habits: This insect is found world-wide. It gets its name from its habit of feeding on the hair padding, feathers, or woolen upholstery of furniture. It also feeds on wool, hair, fur, feathers, bristles, horn, tortoise shell, silk, insects, dried carcasses, dried cheese, dried blood, glue, and book bindings. It damages linen, cotton, jute, softwood, and leather when these items are soiled. Outdoors, the adults feed on plant pollen.

Control: Correct identification of the pest is the initial step in carpet beetle control, followed by a thorough investigation to identify and eliminate the sources of the infestation. The customer should be questioned about past pest problems, birds, flies and other insects, rodents, etc., and the sites of these infestations should be closely examined.

All products containing natural materials should be examined, paying particular attention to stored clothing, carpets, and food products. Sanitation, including removal and/or treatment of infested material is the first step in remediating the problem. Infested food should be discarded. Carpets, rugs, and clothing should be brushed and cleaned.

Application of a residual insecticide to cracks and crevices and the immediate area around an infestation site with an appropriately labeled product might be necessary. In cases where the infestation is associated with a wall void or other inaccessible space, a dust formulation which provides better residual activity should be used.

The exterior of the structure should be inspected for plants that attract the adult beetles, which should either be removed or treated with an appropriately labeled product. During warmer months, perimeter foundation treatment with a wettable powder or microencapsulated formulation reduces the number of beetles entering the structure. Carpet beetle infestations are persistent and often require several follow-up visits in order to resolve the problem.

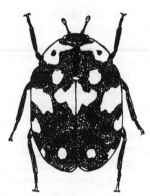

Furniture Carpet Beetle

LARDER BEETLE

Order/Family:
Coleoptera/Dermestidae

Scientific Name:
Dermestes lardarius Linnaeus

Description: Larder beetle adults are 1/4- to 3/8-inch long, elongated-oval, dark brown to black beetles with a distinct clubbed antennae. The upper one-third of the wing covers are covered with a wide, yellow band with six to eight dark spots on it. The underside of the body is covered with fine yellow hairs. The larvae are 3/8- to 5/8-inch long, brown, and very hairy. They have two backward curved, sharp spines on the upper side of the next to the last abdominal segment.

Biology: During her lifetime, the female lays about 100 eggs on food and into cracks and crevices. Eggs hatch in 12 days and bore into the food material they are infesting. The male larvae molt five times and the females molt six times before they pupate. The larvae leave the food material and wander just before the last molt and often bore into wood or other similar materials to pupate. The life cycle, egg to adult, requires 40-50 days.

Habits: Adult larder beetles overwinter in cracks or crevices in outdoor locations and enter buildings in spring and early summer. These insects prefer to infest meats such as ham, bacon, dried beef, dried fish, as well as cheese, feathers, horns, skins, dry pet foods, hair, and museum pieces. In homes these beetles may be found infesting accumulations of dead cluster flies and face flies and dried rodent carcasses. Most damage is done by the larvae; however, the adults also feed. The adults avoid light during mating and egg-laying.

Control: Larder beetles are not as common as they once were because few people now cure meat in their homes. When these insects do appear, the homeowner usually finds the adult beetles or the wandering larvae.

Control of the larder beetle begins with a thorough inspection to discover the source of the infestation. The customer should be questioned about previous rodent, bird, and fly infestations. Locating the source(s) of the infestation might be difficult because larder beetle infestations often are in a wall void or attic where dead insects, birds, or rodents provide a food source for the larvae. Whenever possible, the source of the infestation should be removed. Thorough sanitation in meat packing plants or factories that handle hides, feathers, or other suitable breeding materials is absolutely essential to prevent larder beetle infestations. Residual insecticide sprays or dusts can be applied into cracks and crevices or voids that appear to be sources of these insects.

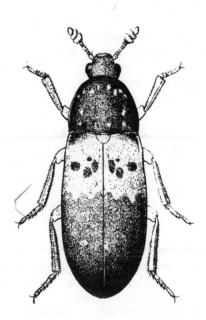

Larder Beetle

RED & CONFUSED FLOUR BEETLES

Order/Family:
Coleoptera/Tenebrionidae

Scientific Name:
Tribolium confusum — (Duval) Confused Flour Beetle
Tribolium castaneum — (Herbst) Red Flour Beetle

Description: The adult beetles are red-brown, slender, and about 1/8-inch long. Both species look very similar but can be distinguished by looking at the antennae. The confused flour beetle's antennae gradually enlarge toward the tip, ending in a four segmented club. The red flour beetle's antennae become club-like very quickly and the club has three segments. In addition, the sides of the confused flour beetle's pronotum (segment right behind the head) are very straight, whereas the red flour beetle has curved sides on its pronotum. The yellowish-white elongate larvae are 1/8- to 1/4-inch long and the last segment of the body has two darkened unsegmented and immovable prongs.

Biology: Over their lifetime of two to three years, females produce 300 to 500 eggs. They lay two to three of these clear, white, sticky eggs daily in cracks, in bags, or through the mesh of flour bags. The eggs hatch in five to 12 days, and the larvae undergo five to 12 molts, completing development in about 30 days. The life cycle, egg to egg, can be completed in 49 to 90 days. Under ideal conditions and at increased temperatures, these insects can have four to five generations per year.

Habits: Confused and red flour beetles are major pests of flour. They cannot feed on whole grains but are found abundantly in grain dust, flour, dried fruit, nuts, chocolate, snuff, spices, rodent baits, and drugs. Flour that is heavily infested with flour beetles has a disagreeable odor and flavor caused by secretions from the insect's scent glands. Confused and red flour beetles are attracted to light; however, only the red flour beetle is reported to fly.

Control: Flour beetles are one of the most common stored products pests found within houses. Infested flour or other stored products should be identified and discarded or heat treated (120° F for several hours). The cupboards, cabinets, closets, shelves, and pantry where the infestation is located should be well vacuumed to eliminate spilled flour and other food dusts. Special care should be taken to remove materials from cracks and crevices. Pheromone traps can be use to detect and monitor the red flour beetle.

Storage areas and sites of infestation should be treated with liquid or dust formulations as a supplement to elimination of the source of the infestation. Infestations in mills, food processing, storage, and manufacturing facilities are controlled with sanitation, residual applications, and, when necessary, fumigation.

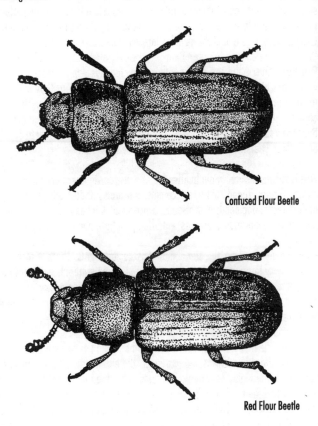

Confused Flour Beetle

Red Flour Beetle

SAW-TOOTHED GRAIN BEETLE

Order/Family:
Coleoptera/Cucjidae

Scientific Name:
Oryzaephilus surinamensis (Linnaeus)

Description: Adult saw-toothed grain beetles are small, slender, dark brown, flat insects about 1/8-inch long. Their most distinguishing characteristic is the six saw-like teeth found on either side of their pronotum (first segment behind the head). The larvae are yellow-white and about 1/8-inch long. They have three pairs of legs and a pair of false legs on the abdomen.

Biology: Over a four to five month period, females deposit 45-285 shiny white eggs in cracks or crevices in the foods they are infesting. They molt two to four times, and the life cycle (egg to egg) usually requires 30-50 days, although it may take as long as 375 days. There can be six to seven generations per year. Adults usually live 6-10 months but have been known to live for up to three years.

Habits: Saw-toothed grain beetles feed on a wide variety of stored products including flour, bread, breakfast cereals, macaroni, dried fruits, nuts, dried meats, sugar, dog food, and biscuits. Since these beetles are very flat, they easily hide in cracks and crevices and often penetrate poorly sealed packaged foods.

The larvae feed on the same materials as the adults. These insects can develop very large populations in seldom-used stored materials, such as flour. The homeowner typically becomes aware of the infestation when the adults are seen crawling actively about the pantry area. The adults are not known to fly and are not attracted to light.

Control: The first step in effective saw-toothed grain beetle control is to find, remove and destroy all infested materials. The entire storage area should be vacuumed thoroughly to remove flour and other food materials from cracks and crevices. These materials can support adults and larvae. Infested stored products can also be sterilized with heat (125°F for one hour) or cold (0-5°F for 24 hours).

The surfaces, cracks and crevices, and other potential harborage areas for these pests can be treated with residual insecticide sprays or dusts. The application of these materials also helps to control adults that crawl into storage areas and serve as potential reinfestation threats.

Sanitation is very important in all grain storage or processing facilities to keep potential breeding sites to a minimum. Heavily infested materials may need to be fumigated to achieve control.

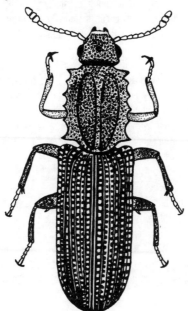

Saw-Toothed Grain Beetle

SPIDER BEETLES

Scientific Name:

Ptinus villiger (Reitter) — Hairy Spider Beetle
Ptinus fur (Linnaeus) — White-marked Spider Beetle

Order/Family:
Coleoptera/Ptinidae

Ptinus clavipes — Panzer Brown Spider Beetle
Ptinus ocellus — Brown Australian Spider Beetle

Description: Spider beetles are small, oval, or cylindrical beetles that are 1/32- to 3/16-inch long. The abdomen is usually globe-like and the prothorax (first segment behind head) is constricted where it attaches to the rest of the body. They have long hairy legs and antennae, and the head is concealed from above by the prothorax, making these beetles look much like spiders or large mites. The larvae are 1/8- to 3/16-inch long, white, and "C" shaped.

The hairy spider beetle is red-brown with two odd-shaped white patches on each wing cover; the body and wing covers are covered with rows of small pits. The white marked spider beetle is also red-brown and covered with yellow hairs; the females have two white patches on each wing cover. The brown spider beetle is uniformly brown. The Australian spider beetle is dark red-brown, and covered with golden brown or yellow hairs that cover rows of pits running the length of the body.

Biology: The adult female lays a few eggs, often less than 100, in or near the larval food. The larvae molt three times before spinning a debris-covered cocoon within which they pupate. These beetles survive best in old buildings, often boring into old wood or cardboard boxes to pupate. The developmental time (egg to adult) varies according to the species: Australian spider beetle (94 days), the brown spider beetle (6-9 months), and the white marked spider beetle (32 days).

Habits: These beetles are general scavengers which feed on a wide variety of plant and animal foods including grain products, seeds, dried fruits or meats, wool, feathers, rodent droppings, and dried insects. The small, white larvae develop on the same foods as the adults. Spider beetles can remain active during freezing weather and are usually a stored product pest in the very northern states. Most species are active only at night, hiding in cracks and crevices during the day. Some species fly and are attracted to lights at night.

Control: Spider beetle infestations can be difficult to control because they can live on such a wide variety of foods and are active mainly at night. Sanitation and rotation are the most important steps in control. The source of the infestation is often found in a wall void, drop ceiling, or similar undisturbed area where food materials have collected. These sites should be cleaned where practical and access to them sealed off if possible. Infested materials should be removed and discarded.

Liquid and dust formulations should be used to treat cracks and crevices and other voids where the beetles might be hiding during the day and to prevent future infestations. Adult populations can be reduced at night by the application of ULV formulations. Fumigation for this pest is rarely required.

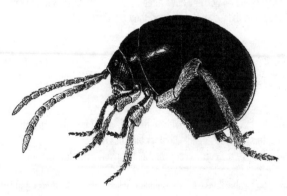

Spider Beetle

VARIED CARPET BEETLE

Order/Family:
Coleoptera/Dermestidae

Scientific Name:
Anthrenus verbasci (L.)

Description: Adult varied carpet beetles are about 1/16- to 1/8-inch long and generally oval in appearance. The wing covers meet at the rear end of the body in a smoothly rounded fashion with no apparent cleft. The back of these insects are spotted with gray-yellow, brown, and white scales which form two bands across the top. The underside of the body is gray to yellow.

The larvae are 3/16- to 1/4-inch long and wider at the end of the body than at the head. They are covered with a series of light and dark-brown stripes that run across the body. They have three dense tufts of bristles at the rear-end of the body that extend outward when they are disturbed.

Biology: Females lay about 40 eggs in a lifetime; eggs hatch in 10 to 20 days. The larvae develop in 222 to 323 days and remain as pupae for 10 to 13 days. There is one generation per year. The adult beetles live 14 to 44 days.

Habits: The larvae feed on a wide variety of foods, including carpets, woolens, skins, furs, stuffed animals, leather bindings on books, feathers, silk, and plant products, (e.g., cacao, corn, and red pepper). The favorite larval food, however, is dried insects or spiders, and, thus, these insects are terrible pests in dried insect collections. Adults feed on pollen, are good fliers, and enter homes through open windows.

Control: The initial step in carpet beetle control is the correct identification of the pest, followed by a thorough investigation to identify and eliminate the sources of the infestation. The customer should be questioned about past pest problems, birds, flies and other insects, rodents, etc. and the sites of these infestations closely examined.

All products containing natural materials should be examined, paying particular attention to stored clothing, carpets, and food products. Sanitation, including removal and/or treatment of infested material, is the first step in remediating the problem. Infested food should be discarded, and carpets, rugs, and clothing brushed or cleaned.

Application of a residual insecticide to cracks and crevices and the immediate area around an infestation site with an appropriately labeled product might be necessary. In cases where the infestation is associated with a wall void or other inaccessible space, a dust formulation which provides better residual activity should be used.

The exterior of the structure should be inspected for plants that attract the adult beetles, and should be removed or treated with an appropriately labeled product. During warmer months, perimeter foundation treatment with a wettable powder or microencapsulated formulation reduces the number of beetles entering the structure. Carpet beetle infestations are persistent and often require several follow-up visits in order to resolve the problem.

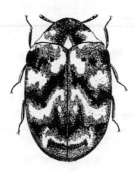

Varied Carpet Beetle

ANGOUMOIS GRAIN MOTH

Order/Family:
Lepidoptera/Gelechiidae

Scientific Name:
Sitotroga cerealella (Olivier)

Description: Adult Angoumois grain moths are small, buff or light brown insects with a wingspan of 1/2- to 5/8-inch. The hind wings are gray and taper to a thumb-like projection at their tips. The forewings and hindwings are fringed with long hairs. The fully developed larvae are about 1/4-inch long and are yellow-white with brown heads.

Biology: Females lay an average of 40 eggs during their lifetime, depositing them on or near grain. The larvae hatch in four to eight days and bore into the kernels where they feed and develop. They undergo three molts in two to three weeks, then pupate within the kernel in a silken cocoon. Adults emerge in about 10 to 14 days. The life cycle from egg to egg requires 35 to 49 days. There are usually four or five generations per year, with as many as 10 to 12 generations per year in heated warehouses.

Habits: Angoumois grain moths primarily attack whole grain. The larvae develop within the undamaged kernels of corn, barley, rye, oats, rice, and other seeds, often infesting the grain before it is harvested. Adult moths do not cause damage. Occasionally the larvae develop in caked grain material. The Angoumois grain moth is also a common household pest which infests decorative Indian corn, popcorn, corn used to fill bean bags, and other whole seeds. The adults are excellent fliers and are sometimes confused with the clothes moth. However, adult Angoumois grain moths are attracted to light, and clothes moths are not.

Control: When Angoumois grain moths are discovered within structures, it is essential that infested grain, seeds, or caked material are found and discarded. Infested stored products should be cleaned up and cupboards, shelves, closets, and pantries carefully removed. Decorative items should be placed in a freezer for several days to kill the larvae and pupae. Adults should be removed with a vacuum or light trap, or a pyrethrin space spray should be applied for quick knockdown. It is rarely necessary to make a pesticide application because the larvae and pupae usually are found inside whole grains.

Infestations of Angoumois grain moth in grain storage and processing facilities can be cantrolled by removing and destroying infested products and by using intensive sanitation and fumigation as required.

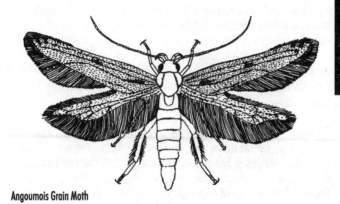

Angoumois Grain Moth

CASEMAKING CLOTHES MOTH

Order/Family:
Lepidoptera/Tineidae

Scientific Name:
Tinea pellionella (Linnaeus)

Description: Adult casemaking clothes moths are approximately 3/8- to 1/2-inch from wing tip to wing tip, and are slightly smaller than webbing clothes moths. The wings and body are buff to golden with a brown tinge and the front wings have three dark spots, but these distinguishing characteristics are often rubbed off.

The larvae are small caterpillars (3/8- inch long) that live within a small portable, silken case which they carry as they feed. The larvae have dark head capsules and the first thoracic segment (leg bearing segment) is dark brown or black.

Biology: Females begin laying eggs (37-48) singly on suitable larval food the day after emergence as an adult. The larva feeds for about 33-90 days and molts 5-11 times. The mature larva then finds a sheltered place to pupate. The insect pupates within the silken larval case. Developmental time (egg to adult) requires 46-116 days. The casemaking clothes moth is usually more common in the southern states where there are two generations per year. Adults may lay eggs year around in the northern states but have only one generation per year.

Habits: The casemaking clothes moth prefers products of animal origin, secondarily feeding on products of plant origin. It is a pest of woolens, rugs, feathers, felts, skins, spices, drugs, furs, taxidermy mounts, and stored tobacco. The larva remains within the case at all times and dies if removed. It can turn completely around without leaving the case and can feed from either end. Adults do not feed.

Casemaking clothes moths shun light, and although the males are active fliers, the females fly only short distances.

Control: The key to controlling this pest is thorough inspection and identification of infested materials. Some unusual locations to look for this pest include air ducts, particularly the cold air return, along baseboards, and other areas where pet hair accumulates.

As with the webbing clothes moth, prevention is the best control. Thorough vacuum cleaning of rugs and furniture removes lint and pet hair as well as some of the larvae. Stored clothing should be kept in tightly closed containers. Spices, tobacco, or drugs should be stored in very tightly closed containers.

Infested rugs, carpets, and furniture should be cleaned thoroughly and protected with a residual insecticide application. Larvae are easily removed from infested clothing by cleaning or laundering. Sensitive items, such as museum pieces, wall mountings, furs, taxidermy mounts, etc., might require vault fumigation or treatment with a dust formulation.

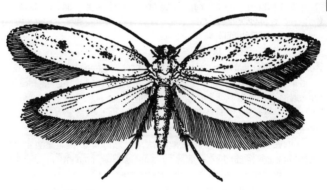

Casemaking Clothes Moth

INDIAN MEAL MOTH

Order/Family:
Lepidoptera/Pyralidae

Scientific Name:
Plodia interpunctella (Hubner)

Description: Indian meal moth adults have a 5/8- to 3/4-inch wing spread. The wings are a burnished copper, almost purple with a broad gray band near where they attach to the body. The mature larvae are about 1/2- inch long and dirty white, pink, brown, or light green. The head and top of the first body segment behind the head are reddish-brown to yellowish brown.

Biology: The adult females lay 100 to 400 eggs at night on the larval food over a one to 18-day period. Larval development requires 13 to 288 days. The average life cycle, egg to egg, requires 25 to 135 days. There are usually four to six generations per year depending on the food supply and temperature. These moths usually overwinter as larvae. The adults are more active at dusk and are attracted to light.

Habits: Indian meal moths are one of the most common stored product pests found in homes, food processing plants, grain storage and processing facilities. The larvae prefer to feed on coarsely ground flour and meal but commonly feed on whole grains, dried fruit, nuts, chocolate, beans, crackers, biscuits, dry dog food, bird seed, and red peppers.

The larvae produce a silk webbing over the surface of the materials upon which they are feeding. The webbing contains large amounts of their frass (feces). The damage caused by this insect's feeding is compounded by the presence of this repulsive mat. The larvae move into cracks and crevices in the food material, feeding within or near this silken mat. The mature larvae often move away from infested materials to pupate in cracks or crevices. This behavior pattern often allows them to be discovered by homeowners.

Control: Inspections and information obtained from the homeowner often lead to the source of the infestation. All products and areas that may be infested, such as abandoned rodent or bird nests, should be inspected for adults, larvae, and webbing. Pheromone traps should be used to identify areas of activity. All infested items should be discarded and all uninfested products should be placed in insect-proof containers or in a refrigerator or freezer.

Shelves, cracks and crevices, and other areas where food products accumulate should be vacuumed and cleaned out. After removing debris, pupae, and webbing, surfaces, cracks and crevices, and other potential harborage areas should be treated with a liquid or dust formulation. Aerosol sprays or mists containing pyrethrins can be used to control adult moths, although this does not solve the infestation problem.

Infestations within food processing plants, grain storage or processing facilities, and warehouses can be controlled using intensive sanitation and fumigation, as necessary.

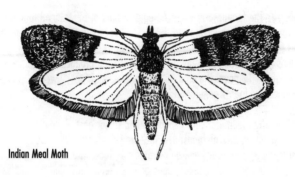

Indian Meal Moth

MEDITERRANEAN FLOUR MOTH

Order/Family:	Scientific Name:
Lepidoptera/Pyralidae	*Anagasta kuhniella* (Zeller)

Description: Adult Mediterranean flour moths are 1/4- to 1/2-long with a wing span of a little less than 1 inch. The forewings are pale gray with wavy black lines running across them. The hindwings are dirty white. When the adult is at rest, it pushes up with its front legs, making the wings look as if they slope downward. The larvae are 5/8- to 3/4-inch long when fully developed. They are white or light pink with a dark head and dark plate just behind the head.

Biology: The females lay 116-678 eggs in flour or other larval food materials where they hatch in three to five days. The larvae feed within the silken tubes they produce, maturing in about 40 days. They pupate within these silken cocoons in clean food material away from infested products. The pupal stage lasts eight to 12 days. The entire life cycle takes eight to 10 weeks. There are four to six (or more) generations per year.

Habits: The Mediterranean flour moth is a major pest of flour mills and infests a great variety of foodstuffs, including nuts, chocolate, seeds, beans, biscuits, flour, dried fruit, and dehydrated dog food. The adults do not cause damage. The larvae spin silken tubes around themselves, and this silk sometimes becomes so thick in flour mills that it clogs machinery, requiring shutdown for cleaning.

The mature larvae are much like Indian meal moths in that they have a tendency to migrate. These larvae are often the first sign the homeowner has that there is an infestation in the home. The adults fly at night in a very characteristic zig-zag pattern and are attracted to light; but they are seldom seen by the homeowner.

Control: When a Mediterranean flour moth infestation is suspected, it is important to find and discard the infested items. These materials usually are found in seldom used flour or cereal products that have ended up in the back of the cupboard. The cupboards, shelves, cabinets, or pantries where the infested material is found should be cleaned carefully with a vacuum cleaner to remove flour and food dust. Insecticide sprays or dusts can be applied to these areas to control larvae that are missed by cleaning. Adult moths can be controlled with aerosol sprays containing pyrehrins.

Sanitation and fumigation are necessary to control this moth in flour mills and other types of food processing facilities.

Mediterranean Flour Moth

WEBBING CLOTHES MOTH

Order/Family:
Lepidoptera/Tineidae

Scientific Name:
Tineola bisselliella (Hummel)

Description: Adult moths are 1/2-inch long from wing tip to wing tip; when the wings are folded, the insect is about 1/4-inch long. The wings and body are buff/golden except for reddish hairs on top of the head. The antennae are darker than the rest of the body, and the eyes are black. The larvae are 1/2-inch long when mature. They are small caterpillars that are a clear to creamy-white color with a light brown head capsule.

Biology: The females carefully place their eggs deep in the mesh of the infested fabric by attaching the eggs with a glue they secrete. Each female lays 40 to 50 eggs which hatch in four days to three weeks. The newly-emerged larvae begin to feed immediately. They often spin silken tunnels or mats, incorporating fragments of the textile being infested and bits of feces into its construction. The larvae molt five to 45 times, depending on conditions, taking from 35 days to two years to finish development. They eventually spin a silken cocoon in which they pupate. Adults live for approximately two weeks.

Habits: Webbing clothes moth larvae feed on clothes, carpets, rugs, upholstered furniture, felt, animal hair, and stored wool. They especially like to feed on soiled materials. Adults have nonfunctional mouthparts and do not feed. These moths are the most common clothes moth found in the United States. Adult webbing clothes moths are seldom seen because they avoid light.

Control: Prevention of infestations by these moths by frequent, thorough cleaning is the most effective control. Thorough vacuum cleaning of rugs and furniture also helps to prevent infestations. Clothing should be stored in tightly closed containers.

Clothing infested with webbing clothes moths should be cleaned or laundered to remove the larvae. Infested rugs and furniture should be thoroughly cleaned and protected with residual insecticides.

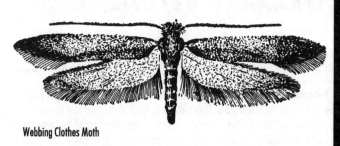

Webbing Clothes Moth

GRANARY WEEVIL

Order/Family:
Coleoptera/Curculionidae

Scientific Name:
Sitophilus granarius (L.)

Description: The adults are polished red-brown to black snout beetles that are 1/8- to 1/4-inch long. They can be distinguished from rice weevils by the absence of four, light-colored spots on the wing covers and the presence of elongate oval pits in the surface of the pronotum (first segment behind the head). The head is elongated into a long, slender snout. The larvae are small, white, legless grubs found within the kernels of wheat, corn, or rice. These grubs have a relatively flat underside of the body, and a rounded upper surface.

Biology: The female bores a small hole in the surface of a kernel of grain, into which she places an egg and covers it with a gelatinous material. Each female is capable of laying more than 200 eggs, and the larvae can develop in all of the commonly stored grains.

The eggs hatch within a few days and the larvae feed within the kernel for 19 to 34 days, molting four times. They pupate within the kernel and emerge as adults five to 16 days later. There are commonly four generations per year.

Habits: Granary weevils are considered to be one of the primary stored product pests because they infest only whole kernels of grain. These insects cannot fly, having lost the use of their second pair of wings; therefore, they are found mostly in grain storage facilities.

Control: Adult granary weevils have a tendency to wander and so may be found far from the source of their infestation. It is important that the infested grain be found and destroyed before implementing any control measures. Whole grain such as popcorn, decorative seed displays, Indian corn, and "bean bags" filled with whole grains are potential sources of infestation. Small quantities of infested grain can be heat-sterilized in the oven. Granary weevil infestations within large storage facilities are successfully controlled with increased sanitation and bulk fumigation of the grain.

To provide control of exposed adults, residual insecticides can be applied to cabinets, closets, and shelves where infested stored grains are found. Care should be taken to apply these materials in all cracks and crevices as well as in any other potential harborage areas.

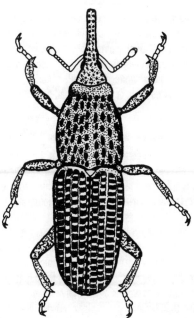

Granary Weevil

RICE WEEVIL

Order/Family:
Coleoptera/Curculionidae

Scientific Name:
Sitophilus oryzae (Linnaeus)

Description: Adult rice weevils are dull red-brown snout beetles that are 1/8-inch long. They have four faint red or yellow spots on their wing covers, and the pits in the pronotum (first segment behind the head) are round or irregularly shaped. The head is elongated into a long slender snout.

The larvae are legless, white grubs with a dark head capsule. They are relatively flat on the underside of their body and are very rounded or curved on the upper side.

Biology: A female rice weevil lays 300-400 eggs in a lifetime. She bores a small hole in a kernel and deposits an egg which she covers with a gelatinous fluid. The larva hatches in a few days and feeds inside the grain kernel, molting four times in 32 days before pupating. The life cycle (egg to egg) takes about a month. The adults live for three to six months.

Habits: Rice weevils are a more common problem in the southern states, although they are found in homes and in stored grain world-wide. Rice weevils are primary stored grain pests, infesting only whole grains. The larvae develop in all the commonly stored domestic grains and the adults have been observed to feed on these grains, as well as on a variety of beans, nuts, cereal products, and some fruits. Unlike granary weevils, adult rice weevils are strong fliers, and, in areas from North Carolina southward, often fly from stored grain to fields of corn, rice, and wheat where they infest grain before harvest and later in storage. In the north, weevil activity is confined to stored grain.

Control: Rice weevil adults in the home indicate the presence of infested whole grain somewhere on the premises. Stored popcorn, dried seed decorations, "bean bags," decorative Indian corn, dried seed-bearing plants, and other stored seeds should be checked as potential sources of infestation. Infested materials should be discarded or the infestation destroyed by extreme heat or cold treatment.

Closets, shelves, cabinets, and pantries should be thoroughly cleaned, after which, a residual product can be applied to the surfaces, as well as into cracks, crevices, and other potential harborage areas. Large storage facilities infested with rice weevils need to increase sanitation and eliminate infestations with fumigation.

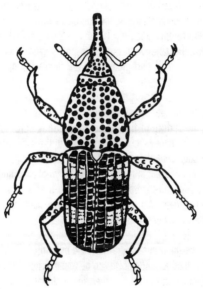

Rice Weevil

ACARID MITES

Order/Family:
Acari/Acaridae

Scientific Names:
Acarus siro Linnaeus — Grain Mite
Tyrolichus casei Oudemans — Cheese Mite

Description: Mites are very small arthropods closely related to ticks which have a broad unsegmented abdomen attached very broadly to the thorax. The adult and nymph stages have eight legs, whereas the immature larval stage has six. The legs have one claw on the end along with a sucker. Cheese mites and grain mites are approximately 1/64-inch long and look identical except under high magnification. Grain mites are pearly white with yellow to red-brown legs. For the most part grain mites are clear with tan mouthparts and legs.

Biology: The female can lay up to 800 eggs in her lifetime. Eggs are deposited singly on the surface of the food material. They hatch into a six-legged larva. The life cycle requires about 15-18 days in the summer and 28 days in the winter. An adult female lives about 40 days.

Grain mites are unusual in that they have a nymphal resting stage called a hypopus which does not feed but is able to attach itself to rodents, birds, or insects with its suckers. As a result of this attachment, this stage is often dispersed on these animals to others areas.

Habits: Both the grain mite and the cheese mite infest grain, cheese, flour, and other stored products. They also can be found in hair-filled upholstery and bedding. Infestation is most likely when these stored products are stored in cool and humid conditions. Heavy acarid mite infestations have a characteristic sweet or minty odor that is readily identifiable. There is a coating of "mite dust" which is actually the molted skins of the mites covering grain sacks or cheeses that are heavily infested. Many times the surface of infested materials appears to move because of the large number of mites present in the infested area.

These mites can cause skin irritation (dermatitis) in humans. This is commonly referred to as grocer's itch or baker's itch. A rash, swelling, and/or small lesions may occur.

Control: Heavy infestations of acarid mites indicate that storage conditions are too damp. Control of moisture and humidity are critical. Products should be stored in a well ventilated and dry location. Maintain the relative humidity below 60%. Materials should be rotated frequently.

Small amounts of infested materials in commercial accounts and homes should be discarded or destroyed and the infested premises thoroughly cleaned. Surfaces in storage areas can be treated with residual liquid or dust formulations. Severely infested grains and cheeses usually require fumigation to achieve control of these pests.

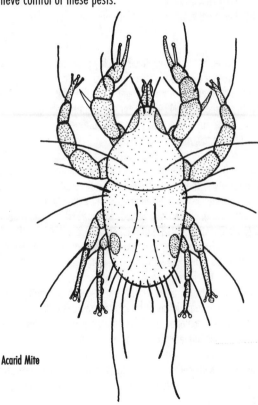

Acarid Mite

CADELLE

Order/Family:
Coleoptera/Tenebrionidae

Scientific Name:
Tenebroides maurintanicus (L.)

Description: Adult cadelles are oblong, shiny black, flattened beetles that are about 1/3-inch long. They look much like ground beetles and are one of the largest insects attacking stored products. The larvae are gray-white and about 1/3-inch long with black head. They have a black plate on the upper side of the last abdominal segment from which two blunt, horn-like projections emerge.

Biology: The females are very long-lived and lay more than 1,000 eggs in batches of 10 to 60 during their lifetime. The eggs are deposited singly or in groups in grain and other foodstuffs and under the flaps of cartons or packages. The larvae require 38 to 414 days for development and pupation lasts for eight to 25 days. There are usually two to three generations per year. Adults can live up to three years.

Habits: Both adults and larvae are capable of penetrating packaging. They feed on a wide variety of foods, including grains, processed cereals, breakfast food, potatoes, and shelled and unshelled nuts. They may burrow into woodwork where they can remain for long periods of time. Adults readily feed on other insects found infesting stored products. Adults are not attracted to light.

Control: A continuous, complete sanitation program is a very important component of cadelle control because removal of grain dust, broken grain, and other infested material also removes a large number of the larvae. For control to be effective, infested material must be located and destroyed. It is important to inspect packaged material, as well as stored products in opened packages, for entrance holes made by the larvae and adults.

These insects feed on whole grain and nuts as well as on processed materials. The presence of adult cadelles might indicate that other stored product pests are present, since the adults feed on these insects. Woodwork in cabinets and pantries should be carefully examined.

Cadelle infestations can be controlled in homes by removing the infested materials and treating the surfaces, cracks and crevices, and other harborage areas in cupboards, cabinets, pantries, and shelves with residual insecticides. Woodwork with signs of tunneling by adults and larvae should be replaced if excessively damaged or sprayed with a residual insecticide.

Cadelle infestations in grain storage and processing facilities are best handled through sanitation and fumigation. However, cadelles may require a higher fumigant concentration or an extended fumigation time. Cold exposure destroys the eggs and pupae; however, adults and larvae can tolerate temperatures of 15^0 to 20^0 F for several weeks.

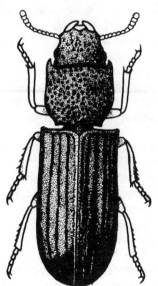

Cadelle

MEAL WORMS

Order/Family:
Lepidoptera/Tenebrionidae

Scientific Name:
Tenebrio molitor Linnaeus — Yellow Mealworm
Tenebrio obscurus Fabricius — Dark Mealworm

Description: Adult meal worms are beetles. Both species are about 1/2- to 5/8-inch long and very similar in shape and color. However, the dark meal-worm adult is a dull black, whereas the yellow meal worm adult is a polished and shiny dark brown or black.

The larvae are 1 1/4-inches long when fully developed. They are cylindri-cal, hard-bodied insects very similar to wireworms. The ninth abdominal seg-ment has a pair of upturned, dark, hard spines on the upper surface. The yellow meal worm larva is bright yellow, while the dark mealworm is dark brown.

Biology: Meal worms overwinter as larvae, emerging in the spring as adults that soon begin laying eggs. During their two to three month life, the adult females lay approximately 276 bean shaped, white, sticky eggs in food mate-rial. The larvae molt 14 to 15 times, over a period of four to 18 months, before pupating near the surface of the food upon which they are feeding. There is one generation per year.

Habits: Meal worms are seldom major stored product pests in the home. The larvae are found in dark undisturbed locations feeding on collections of grain, grain dust and cereal products. Poor sanitary conditions and high mois-ture content are conducive to infestation. Meal worms are medically impor-tant because the eggs and larvae are often eaten in cereals or other breakfast foods and can cause significant gastrointestinal irritation. The adults of both species have well developed wings and are attracted to lights.

Control: The first step in eliminating meal worm infestations is to locate infested material. The larvae may be developing in the corners of cabinets, cupboards, or pantries or in straw-filled furniture. In granaries and warehouses, the larvae may be found in old or discarded grain. The larvae frequently wan-der away from the food in which they are developing, and, therefore, a thorough search should be conducted to identify the source of the infestation.

Infested material may be treated with fumigants, heat or cold. Areas where infestations are discovered can be treated with residual products to control any adults or larvae missed during cleanup.

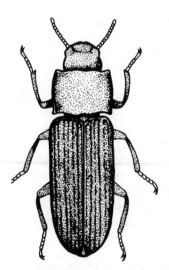

Meal Worm

SILVERFISH

Order/Family:
Thysanura/Lepismatidae

Scientific Name:
Lepisma saccharina Linnaeus

Description: Silverfish are primitive (i.e., older than cockroaches), wingless insects that are 1/2-inch long when fully grown. They are covered with silvery scales and are flattened and somewhat "carrot" shaped. Three long, slender "antennae-like" appendages project from the end of the abdomen, giving them the name "bristle tails."

Biology: The female lays one to three eggs per day in crevices or under objects. The female molts after laying a batch eggs and sheds her skin as many as 50 times after becoming an adult. The eggs hatch in about 43 days at 72-90° F and at least 50-75% relative humidity. The young silverfish look exactly like the adults, except smaller, and feed on the same foods. Under ideal conditions, they molt every two to three weeks becoming adults in three to four months. However, under poor conditions, this might require two to three years. These insects are very long-lived, commonly living at least three years. The silverfish are unlike most other insects in that they continue to molt after they become adults.

Habits: Silverfish are tropical insects that easily adapt to the structural environment. They live in warm (71-90° F), moist locations in the structures; hide during the day; and rest in tight cracks and crevices. They roam great distances looking for food, but once a food source is located, they remain close until the supply is exhausted. They can be found throughout a structure from the basement to individual floors to attics to shingles on the roof. Outdoors, they can be found in mulch, and under siding and roof shingles, particularly cedar shakes.

They readily feed on books, cloth, and sometimes dried meats or dead insects. They seem to be especially fond of the sizing on books and paper, and the glues and pastes found on wallpaper, labels, and paper products.

Control: During the inspection, look for activity in areas that provide moisture, harborage and food. Reducing moisture, lowering the temperature, and removing infested items can help eliminate localized infestations. Sanitation is helpful but may not greatly reduce the problem because these pests feed on so many paper products. They can survive for weeks without food and water.

Silverfish are easily controlled with careful and thorough applications of baits and/or residual insecticide sprays and dusts. Care should be taken to treat wall voids, cracks and crevices, and other suspected harborage areas thoroughly. The same careful application of residual insecticides used for good cockroach control is needed to effectively control silverfish. Outdoors, microencapsulated or wettable powder formulations should be used.

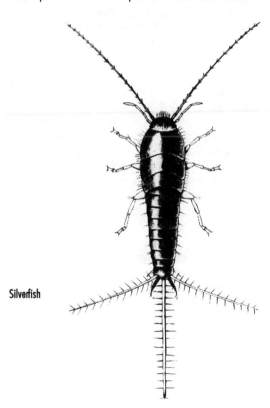

Silverfish

ANOBIID POWDERPOST BEETLES

Order/Family:
Coleoptera/Anobiidae

Scientific Name:
Various

Description: Various anobiid beetles attack seasoned wood in the United States. These beetles range in size from 1/32- to 3/8-inch long; however, those that attack structures are 1/8- to 1/4-inch long. They have highly variable body forms but most are elongate and cylindrical. The first body segment (pronotum) is hood-like, hiding the head when viewed from above. The last three segments of the antenna are lengthened and expanded into a club. The mature larvae are as large as 1/2-inch, C-shaped, dirty white, and the area behind the head is expanded and swollen. The last spiracle on the abdomen is not enlarged.

The furniture beetle, *Anobium punctatum*, is 1/8- to 1/4-inch long, cylindrical, and red-brown to dark brown in color. It has a series of pits in rows that run lengthwise on the wing covers. The pits can be seen through the fine yellow hairs that cover the body. The last three segments of the antenna are longer than the first eight combined.

The adult deathwatch beetle, *Xestobium rufovillosum*, is 1/4- to 3/8-inch long and is gray-brown with patches of pale hairs on the back of the body. It does not have the rows of pits on the wing covers and their 11-segmented antenna end in three elongated segments that are as long as the previous five segments.

Biology: Furniture beetle adults emerge in the spring from cells just below the surface of the infested wood. Soon afterward, mating occurs, and egg laying begins. The female lays 20-60 eggs in old emergence holes or cracks and crevices in the wood. Eggs hatch in six to 10 days. The larvae feed for about one year before pupating for two to three weeks. The wood moisture content required for larval development is 13-30%. When development is complete, the adult bores directly to the surface of the wood, emerging through a round hole 1/16- to 1/8-inch in diameter. Development under ideal conditions can be completed in one year; however, two to three years is more common. The adults are active at night. Some species are attracted to light.

Habits: These beetles commonly infest seasoned sapwood of hardwoods and softwoods; they are rarely found in heartwood. They attack structural timbers, lumber, cabinets, and furniture. These beetles reinfest, and the females commonly lay eggs in the wood from which they emerged. The larvae typically follow the grain of the wood when feeding and fill their tunnels with wood frass. The frass is a fine powder with long pellets loosely packed into the galleries.

Control: Determine if the infestation is active before initiating treatment. Wood in structures and furniture infested by these beetles may go unnoticed until the round adult emergence holes appear in the surface. The characteristic pellets found in the frass and the consistency of the frass are useful in determining what species is infesting the wood. Infested wood can be removed and replaced with treated wood. Reducing the wood moisture content to approximately 12% slows the development of the larvae.

Galleries of existing infestations can be injected with aerosol or dust formulations. The surface of unpainted or otherwise unprotected wood can be treated and the galleries injected with disodium octaborate tetrahydrate. This kills exposed larvae and prevents reinfestation when the eggs hatch and immature larvae begin to penetrate the wood. However, the most effective way to eliminate anobiid powderpost beetle infestations is to fumigate using sulfuryl fluoride or methyl bromide.

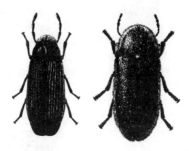

Anobiid Powderpost Beetle

BOSTRICHID BEETLES

Order/Family:	Scientific Name:
Coleoptera/Bostrichidae	Various

Description: The bostrichid beetles range from 1/16- to 1-inch in length. They have an enlarged prothorax (first body segment) which gives them a humpbacked appearance. The body is elongate and cylindrical. The head usually points straight down and is hidden from view by the large prothorax. Most bostrichids have a roughened thorax, and their short antennae usually end in three or four enlarged segments.

The black polycaon, *Polycaon stoutii*, is unlike the majority of the bostrichids. It is 1/4- to 7/8-inch long, coal-black with a prominent head that points straight out from the thorax, and has an oval prothorax with rasp-like teeth. The wing covers have small holes.

The lead cable borer, *Scobicia declivis*, is a more typical bostrichid. It is 1/4-inch long, cylindrical, brown, black, or red-brown beetle with red mouthparts, legs, and antennae. The last three segments of the eight-segmented antennae are greatly enlarged. The head is concealed by the hood-like prothorax which has many puncture-like holes in the front half. The wing covers have deep punctures arranged in rows.

Biology: The bostrichid female differs from anobiid and lyctid beetles because they bore into wood to form an egg gallery where the eggs are inserted into pores in the wood as she moves in and out of the egg tunnel. The eggs hatch in about three weeks, and the larvae feed on the wood for the next nine months. The pupal stage lasts two weeks; however, the new adults stay in the tunnel for four to six weeks, then chew their way out of the gallery, emerging through a round exit hole 1/8- to 1/4-inch in diameter. The developmental time (egg to adult) requires approximately one year.

Habits: The bostrichids attack seasoning or newly-seasoned sapwood of hardwoods less than ten years old. They occasionally attack softwoods. The black polycaon is found primarily in the far Southwest and Pacific Coast states and attacks hardwoods and plywood. The leadcable borer is common along the

Pacific Coast and bores through lead cable coverings, plastic pipes, etc. Bostrichids infest the prunings from fruit trees and ornamentals. Bostrichids rarely reinfest. The larvae bore parallel to the grain, filling the tunnels with tightly packed meal-like frass which tends to stick together.

Control: The first indications of bostrichids within a structure is usually the emergence holes formed by the adults when they leave the larval galleries and frass which accumulates beneath the infested material. Infested wood can be removed and replaced with treated wood.

Galleries of existing infestations can be injected with aerosol or dust formulations. The surface of unpainted or otherwise unprotected wood can be treated and the galleries injected with disodium octaborate tetrahydrate. This kills exposed larvae and prevents reinfestation when the eggs hatch and immature larvae begin to penetrate the wood. The most effective strategy to eliminate bostrichid beetle infestations is to fumigate using sulfuryl fluoride or methyl bromide. However, fumigation offers no residual protection. From the consumer's perspective convenience and expense often determine the method of control.

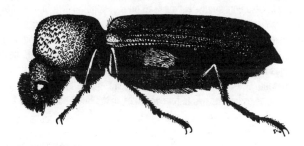

Bostrichid Beetle

LYCTID POWDERPOST BEETLES

Scientific Name:

	Lyctus planicollis (LeConte) — Southern Lyctus
	Lyctus cavicollis (LeConte) — Western Lyctus Beetle
Order/Family:	*Lyctus brunneus* (Stephens) — Brown Lyctus Beetle
Coleoptera/Lyctidae	*Trogoxylon parallelopipedum* — Velvety Powderpost Beetle

Description: There are several other species in the family Lyctidae in addition to those named above that infest seasoned hardwoods. Adult lyctid beetles range in size from 1/32- to 1/4-inch long. They are red-brown to brown or black in color, and their prominent head is easily seen from above. The last two segments of their 11-segmented antennae are expanded into a club. The tibiae, which are the fourth leg segments, have prominent spurs.

The larvae are tiny "C" shaped grub-like larvae found feeding in tunnels in the wood. They are usually less than 1/4-inch long, with an enlarged first body segment (prothorax) and eight spiracles (breathing holes) in the abdomen. The last spiracle is very large compared to the others. These larvae have three-segmented antennae and three-segmented legs.

Biology: Adult females lay 15 to 50 eggs soon after mating deep within the pores of these hardwoods. Larvae feed on the wood for two to nine months, then they migrate to near the wood surface and pupate. The development time depends on the wood's starch content. They do not infest wood with less than three percent starch content. When development is complete, the adult emerges through a round hole, 1/32- to 1/16-inch in diameter, in the surface of the wood. Development (egg to adult) usually requires nine to twelve months but it can be as long as two to four years, or, in the South, as short as three to four months.

Habits: Lyctid powderpost beetles infest the sapwood of seasoned hardwoods including oak, hickory, ash, mahogany, and bamboo. The eggs are never deposited on polished, waxed, varnished, and painted surfaces, so damaged items with these type finishes must have been infested before they were finished and/or the eggs were laid on an unfinished surface, such as the underside of a table or chair. They can reinfest the same piece of wood until it is reduced to a shot-hole riddled shell filled with frass the consistency of face powder. The wood moisture content suitable for larval development is 8 to 32% but is ideal at 10 to 20%. Adults are active at night and are attracted to light.

Control: Prior to initiating any treatment it should be determined if the infestation is active. Infestations are first noticed when small, "bird shot" size holes appear in the wood and very fine powder-like frass falls from the holes and accumulates in piles under the infested item.

If practical, infested wood can be removed and replaced. Small items not susceptible to cold or moisture damage can be placed in a freezer for several days at 0° F. Infestations in lumber can be eliminated by fumigation or kiln drying and prevented by treating raw lumber with protective sprays. The lumber used in floors or furniture can be protected with varnishes, waxes, and paints.

Galleries in wood with existing infestations can be treated with aerosols or dusts. Fumigation with sulfuryl fluoride or methyl bromide is the most effective means of eliminating infestations of lyctid powderpost beetles in structures and furniture because these gases penetrate inaccessible areas. Sulfuryl fluoride must be used at higher rates to kill lyctid beetle eggs. Heat treatment has also been used to treat entire structures. Heat and fumigants provide no residual protection, and, therefore, unprotected hardwoods may be reinfested unless the wood is treated with a preservative or is finished using paint or varnish.

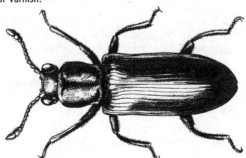

Lyctid Powderpost Beetle

OLD HOUSE BORER

Order/Family:
Coleoptera/Cerambycidae

Scientific Name:
Hylotrupes bajulus (Linnaeus)

Description: Old house borers are members of a large beetle family called the long-horned beetles, named because of their long antennae which are about one-third the body length (or longer). The adult insects are large beetles, approximately 5/8- to 1-inch long. Adult old house borers are slightly flattened, brownish-black with gray or yellow-gray hairs on the upper surface of the body which may be rubbed off the older beetles. The first body segment behind the head (prothorax) has two raised, black, shiny knobs on the upper surface that look like a pair of eyes. The hard wing covers have two wavy, light colored lines that cross them from side to side about one-third of the way down their length.

Old house borer larvae are worm-like, creamy-white, and up to 1 1/4-inches long when mature. They have a broad prothorax which tapers spindle-like to the abdomen. The abdominal segments have deep folds between them so that the abdomen looks as if it is composed of a series of large beads. This stage has prominent chewing mouthparts and very tiny legs on the first body segments (thorax). Old house borer larvae have three simple black eyes (ocelli) in a row at the base of each antenna.

Biology: Adults emerge in early summer to mate, and the females lay 40 to 50 eggs in checks, cracks, or crevices of suitable wood. The hatched larvae bore into the wood and develop during the next two to 10 years with the majority taking from three to five years to reach the pupal stage. The pupal stage lasts only about two weeks, but the new adults remain in the tunnels for seven to 10 months before emerging through oval holes which are 3/8-inch at their widest diameter. Adults live approximately 16 days.

Habits: Old house borers can infest old and new houses. They are the only long-honed beetles that infest seasoned lumber. Old house borer larvae feed only on coniferous lumber such as pine, spruce, fir, and hemlock. The rasping and ticking of the larval feeding activity often is heard by building-occupants and is one of the first signs of an infestation. The galleries are filled with fine frass mixed with small bun-shaped pellets. The surface of infested wood often has a wavy, blistered appearance.

Control: Old house borer infestations are discovered by hearing the larvae feeding in the wood, finding larvae in the wood, seeing accumulating frass, discovering the emergence holes, and seeing adult beetles. Determine if the infestation is active and then recommend treatment. Infested wood can be removed and replaced with treated wood. Reducing the wood moisture content to less than 10% slows the development of the larvae and often kills them.

Galleries of existing infestations can be injected with aerosol or dust formulations. The surface of unpainted or otherwise unprotected wood can be treated and the galleries injected with disodium octaborate tetrahydrate. This kills exposed larvae and prevents reinfestation when the eggs hatch and immature larvae begin to penetrate the wood. The most effective way of controlling old house borer infestations is by fumigation using sulfuryl fluoride or methyl bromide. Fumigation, however, offers no residual protection.

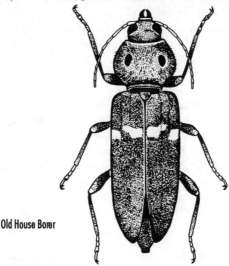

Old House Borer

DAMPWOOD TERMITES

Order/Family:
Isoptera/Hodotermitidae/
Kalotermitidae/Rhinotermitidae

Scientific Names:
Zootermopsis angusticollis (Hagen) —
Common Dampwood Termite
Zootermopsis nevadensis (Hagen) —
Small or Dark Dampwood Termite

Description: These dampwood species are found in states west of the Rocky Mountains. A similar termite species is found in South Florida and the Caribbean. Termites are social insects which live in large colonies. There are three castes: reproductives, workers, and soldiers. Termite antennae have bead-like segments. The winged reproductives (swarmers) have a pair of long wings (equal in size) attached to the last two thoracic segments. The wings are broken off after swarming. The abdomen is broadly joined at the thorax unlike the narrow abdominal attachment found on ants.

Dampwood termite winged reproductives are light brown with dark brown, leathery wings which can be up to one inch long. Some species have a fontanelle (i.e., a small opening on their heads) and/or ocelli (i.e., simple eyes). The nymphs, or workers, are 1/8- to 1/3-inch long and white to cream in color with dark abdomens. The soldiers are 3/8- to 3/4-inch long with large heads with long black-toothed mandibles. Their heads are black at the front and change gradually to a red brown, and their bodies are light brown. Fecal pellets are oval with flat sides, about 1/32-inch long and the color of the infested wood.

Biology: The southeastern species swarms in the spring; the desert and western species swarm in summer through fall. Most species swarm in the evening and are attracted to light. At most, dampwood termites produce a few hundred swarmers compared to subterranean termites which produce thousands of swarmers. The founding male and female create a sealed chamber in the wood and within two weeks produce approximately 12 eggs. The second batch of eggs is laid the following spring. The colonies gradually increase in size, some reaching as many as 4,000 individuals.

Habits: Except for the desert dampwood termite, dampwood termites do not require soil-to-wood contact, but the wood they infest must have a very high moisture content. These termites infest almost any kind of wood and tolerate very high levels of moisture. They often are found in dock and wharf pilings, logs, stumps and dead trees.

As they excavate wood while feeding, dampwood termites litter the inside of their tunnels with their fecal pellets. Dampwood termites usually enter structures in areas where there is wood soil contact and water damaged wood.

Control: Infested structures should be inspected for wood-soil contact, areas where there is a constant water supply such as a plumbing leak, damaged wood, and fecal pellets. The most effective methods of controlling dampwood termites are eliminating the moisture from the wood and removing infested wood and replacing it with treated lumber. Wood remaining in place and galleries can be treated with residual insecticides such as disodium octaborate tetrahydrate which is particularly effective when the wood has a high moisture content.

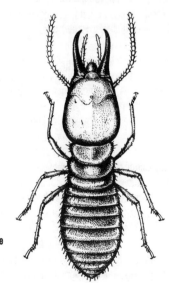

Dampwood Termite

DRYWOOD TERMITES

Order/Family:
Isoptera/Kalotermitidae

Scientific Names:
Incisitermes minor (Hagen) —
Western Drywood Termite
Incisitermes snyderi (Light) —
Southeastern Drywood Termite

Description: Termites are social insects which live in large colonies. There are three castes: reproductives, workers, and soldiers. Termite antennae have bead-like segments. The winged reproductives (swarmers) have a pair of long wings (equal in size) attached to the last two thoracic segments. The wings are broken off after swarming. The abdomen is broadly joined at the thorax unlike the narrow abdominal attachment found on ants.

Drywood termite swarmers are light yellowish brown and the wings have three or more dark veins on the leading edge. They are 3/8- to 5/8-inch long and have red-brown heads and thoraxes. The nymphs are creamy-white with a yellow-brown head. The soldiers have large, parallel-sided, red-brown heads with massive mandibles which have an unequal number of teeth. Their bodies are light colored. Fecal pellets are hard, elongate-oval in shape, 1/25-inch long, and have blunt ends and six concave sides.

Biology: Drywood termites are a nonsubterranean species, i.e., they neither live in the ground nor maintain contact with the soil; they do not build mud tubes. Depending on the species, swarming occurs anytime, spring through autumn. Most species swarm at night and are attracted to light. The male and female swarmer reproductives locate a crack or knothole in a suitable piece of wood; gnaw a small tunnel; seal the entry; excavate a small gallery; and mate. The first year, they produce a few eggs, one soldier, and 20 nymphs. In subsequent years, the colony slowly builds until after 15 years a mature colony could have 3,000 individuals, much smaller than subterranean termite colonies. There is no true worker caste; the nymphs perform their tasks for the colony. After seven molts, the nymphs become either adult reproductives or soldiers.

Habits: Drywood termites are found primarily in the southern United States including the Southwest. They are occasionally found in northern states when the wood (e.g., furniture) they are infesting is moved into these areas. Colonies do not survive in unheated structures. Drywood termites can establish a colony in totally non-decayed wood and maintain residence as long as the wood lasts. Colonies are found in various locations, e.g., attics, door and window frames, trim, eaves, and furniture.

Drywood termites feed across the grain of the wood, creating chambers and/or galleries which are very clean and appear to be sanded smooth. The first signs of infestation are swarmers or the accumulation of fecal pellets below "kick out" holes.

Control: During the inspection, signs of infestation, particularly in attics, should be identified. Occasionally, infested wood can be removed and replaced. If reinfestation is a concern, soffit and ridge vents should be screened in order to prevent reentry of swarmers and exposed wood should be treated with disodium octaborate tetrahydrate.

Drywood termites are most effectively controlled using a fumigant, either sulfuryl fluoride or methyl bromide. This involves tenting the structure and maintaining the concentration of the fumigant for a predetermined amount of time. Heat treatment is a relatively new strategy which requires tenting the structure and achieving and maintaining an interior wood temperature of 120° F for 30 minutes. Some items may be damaged by higher core temperatures in the structure.

A number of spot treatments, such as liquid nitrogen (cold), microwaves, electricity, and applications of residual insecticides are available. The success of these treatments depends on locating the specific area of infestation.

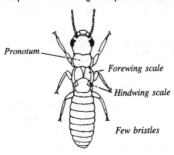

Pronotum

Forewing scale

Hindwing scale

Few bristles

Drywood Termite

FORMOSAN TERMITE

Order/Family:
Isoptera/Rhinotermitidae

Scientific Names:
Coptoternes formosanus Shiraki

Description: Formosan termites have been found in the Gulf Coast states, along the eastern seaboard north to North Carolina, as well as in Tennessee, California, and Hawaii.

Termites are social insects which live in large colonies. There are three castes: reproductives, workers, and soldiers. Termite antennae have bead-like segments. The winged reproductives (i.e., swarmers) have a pair of long, equally-sized wings attached to the last two thoracic segments. The wings are broken off after swarming. The abdomen is broadly joined at the thorax unlike the narrow abdominal attachment found on ants.

Formosan termites are imported pests that are 1/2-inch long, larger than the native subterranean species. The winged reproductives are pale yellow to brownish yellow, and the hairy wings have two dark veins at the leading edge. They have a small pore (i.e., fontanelle) on the front of their heads. The soldiers are easily distinguished from other subterranean termite soldiers because they have an oval head with massive toothless mandibles which cross at the tips.

Biology: Formosan termites are subterranean termites which usually live in the ground, build mud tubes, and construct carton nests, consisting of soil and wood cemented together with saliva and feces. Swarms appear on warm and rainy days around dusk and continue into the evening. The swarmers are attracted to light.

A mature queen produces 1,000 eggs per day. An average colony contains 350,000 individuals, but colonies which contain millions of individuals are not uncommon.

Habits: This is the most destructive termite species which infests structures within the United States. These termites are very industrious, easily out-competing the native species in tunnel- building and destruction of wood. An average size colony consumes 14 linear feet of pine studs in one year. In three months they can cause extensive damage to wood in structures. Their carton nests retain moisture and enable colonies and satellite colonies to establish aerial nests and survive without maintaining contact with the soil. Workers in a colony may forage over an area more than an acre in size and travel almost 400 feet to a food source.

Control: Control of Formosan termites begins with a thorough inspection to determine the extent of the infestation, the points of entry into the structure, moisture problems and other contributing factors, and the development of a control plan. The initial step in subterranean termite control is to remove any scrap wood, firewood, or any other wood materials in contact with the soil. There should be at least six inches between wood and/or foam insulation and exterior soil grade and 18 inches between soil in a crawl space and the floor joists. Water leaks and situations where there is more than 15% wood moisture content must be corrected. Drainage around the structure should be designed so that there is no accumulation of water near the foundation. Situations contributing to this condition include poor exterior soil grades, planter boxes, sprinkler systems, air conditioner condensate drains, lack of gutters, gutter downspouts, etc.

Mechanical alterations and removal of infested wood and replacement with treated wood is useful, but it does not usually stop the termites from gaining access to wood in the structure.

Formosan termites are controlled in the same manner as other subterranean termites - establish a repellent termiticide barrier in the soil and prevent the workers from foraging for food from the colony in the soil to wood in the structure. Soil treatments do not last indefinitely and any disturbance of the soil barrier can result in reinfestation. The recent development of termite baits offers an option for eliminating the colony using an insect growth regulator or slow acting toxicant, but this technique does not immediately stop termite foraging activity within the structure.

Carton nests should be removed from the structure, however, if this is not possible, they should be treated with residual insecticides. The best method for destroying aerial colonies is fumigation with sulfuryl fluoride or methyl bromide. This involves tenting the structure and maintaining the concentration of the fumigant for a predetermined amount of time. Heat treatment is a relatively new strategy which requires tenting the structure and achieving and maintaining an interior wood temperature of 120° F for 30 minutes. Some items may be damaged by higher core temperatures in the structure.

A number of spot treatments, such as liquid nitrogen (cold), microwaves, electricity, and applications of residual insecticides are available. The success of these treatments depends on locating the specific area of infestation. Treatment with disodium octaborate tetrahydrate kills the termites and protects treated wood from attack.

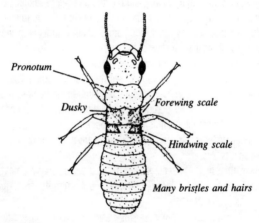

Pronotum

Dusky

Forewing scale

Hindwing scale

Many bristles and hairs

Formosan Termite

SUBTERRANEAN TERMITES

Scientific Names:
Reticulitermes flavipes Kollar —
Eastern Subterranean Termite

Order/Family:
Isoptera/Rhinotermitidae

Reticulitermes hesperus Banks —
Western Subterranean Termite

Description: Eastern subterranean termites are found from Ontario southward and from the eastern United States seaboard as far west as Mexico, Arizona and Utah. Western subterranean termites are found along the Pacific Coast to western Mexico and east into Idaho and Nevada.

Termites are social insects which live in large colonies. There are three castes: reproductives, workers, and soldiers. Termite antennae have bead-like segments. The winged reproductives (i.e., swarmers) have a pair of equally-sized long wings that are attached to the last two thoracic segments. The wings break off after swarming. The abdomen is broadly joined at the thorax unlike the narrow abdominal attachment found on ants.

The winged reproductives are dark brown to almost black and about 3/8-inch long. The wings are brownish gray with a few hairs and two dark veins on the leading edge. They have a very small pore (i.e., fontanelle) on their heads. The soldiers are wingless with white bodies, rectangular yellow-brown heads which are two times longer than their width, and large mandibles which lack teeth.

Biology: Subterranean termite colonies usually are located in the soil from which the workers build mud tubes to structural wood where they then feed. Subterranean termite colonies are always connected to the soil and/or close to a moisture source.

Termites digest cellulose in wood with the aid of special organisms within their digestive system. The workers prefer to feed on fungus-infected wood but readily feed on undamaged wood as well. The foraging workers feed immature workers, reproductives, and soldiers with food materials from their mouths and anuses.

A mature queen produces 5,000 to 10,000 eggs per year. An average colony consists of 60,000 to 250,000 individuals but colonies numbering in the millions are possible. A queen might live for up to 30 years and workers as long as five years.

Habits: Subterranean termite colonies are established by winged reproductives which usually appear in the spring. Swarms usually occur in the morning after a warm rain. A male and female that have swarmed from an established colony lose their wings and seek a dark cavity inside which they mate and raise the first group of workers. Both of these reproductives feed on wood, tend the eggs, and build the initial nest.

After the workers mature, they take over expanding the colony and feeding the reproductives. As the colony becomes larger, light colored supplementary reproductives are produced to lay eggs which then become workers. The soldiers, which are also produced as the colony increases in size, are responsible for repelling invading ants and other predators.

Control: Subterranean termite infestations may go unnoticed until the winged reproductives "swarm" from or inside the structure. The presence of swarmers is a good sign that a well-established colony is in the house and/or its immediate vicinity. Other evidence of infestation is wood damage, i.e., the spring, or soft, wood is eaten out leaving paper-thin walls between the galleries, and the presence of mud tubes. Control of subterranean termites begins with a thorough inspection to determine the extent of the infestation, the points of entry into the structure, moisture problems and other contributing factors, and the development of a plan of control. In addition, it is necessary to sound structural timbers with a screwdriver or other sharp instrument in order to find galleries which have been excavated inside the wood.

The first step in subterranean termite control is to remove scrap wood, firewood, and any other wood materials that are in contact with the soil. There should be at least six inches between wood and/or foam insulation and exterior soil grade and 18-inches between soil in a crawl space and the floor joists.

Water leaks and situations in which there is more than 15% wood moisture content must be repaired. Drainage around the structure should be designed so that there is no accumulation of water near the foundation. Situations contributing to this condition include poor exterior soil grades, planter boxes, sprinkler systems, air conditioner condensate drains, lack of gutters, gutter downspouts, etc. Mechanical alterations, removal of infested wood, and replacement with treated wood are useful measures, but they are not usually successful in stopping the termites from gaining access to wood in the structure.

Traditionally the overriding goal in subterranean termite control has been to provide a barrier between the colony in the soil and the wood in the structure to be protected. This barrier can be achieved with mechanical alterations of the structure, termite-proof building materials, and insecticidal barriers. Most subterranean termite control jobs use a combination of these methods, but soil treatment with residual insecticides is the most common control or preventive method. Soil treatments do not last indefinitely, and any disturbance of the soil barrier can result in reinfestation. The development of termite baits offers an option for eliminating the colony using an insect growth regulator or slow acting toxicant but this technique does not immediately stop termite-foraging activity within the structure.

Galleries excavated in wood can be injected with aerosol residual insecticides to achieve temporary control of workers and winged reproductives in these localized areas. Treatment with disodium octaborate tetrahydrate kills the termites and protects treated wood from attack.

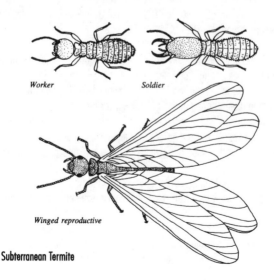

Worker *Soldier*

Winged reproductive

Subterranean Termite

CARPENTER BEES

Order/Family:
Hymenoptera/Anthophoridae

Scientific Name:
Xylocopa spp.

Description: Carpenter bees are large (1/2- to 1-inch long), robust insects that look like bumble bees. They differ by having a bare, shiny black abdomen compared to bumble bees which have a hairy abdomen with some yellow markings. Male carpenter bees, identified by the bright yellow spot in the middle of the head, are aggressive but quite harmless since they lack stingers. Females can sting if molested.

Biology: Adults overwinter in galleries, emerging in the Spring to mate. The female prepares a nest by excavating a new site or more frequently by cleaning out and expanding an existing tunnel. After the nest is ready, she places a mass of pollen mixed with nectar in the blind end of the tunnel, lays an egg on it, and builds a partition of chewed wood pulp to form a brood cell. Six to eight brood cells are constructed in each tunnel. The bee larvae develop on the pollen and emerge as adults 30 to 40 days later, usually in late summer. There is one generation per year.

Habits: Carpenter bees actually bore holes into wood to create a tunnel in which to raise their young. Carpenter bees are not social insects, i.e., they do not live in nests or colonies like yellow jackets and honey bees. The entry hole is 3/8- to 1/2-inch in diameter and initially about 6-inches long; in subsequent years, however, this maybe extended to more than ten feet. The initial opening is straight into the wood, then the gallery typically makes an abrupt right angle and follows the grain of the wood and parallel to the outer surface. Entry holes are usually located in well-lit and sheltered areas, such as headers, roof eaves, porch ceilings, fascia boards, decks, doors, and window sills. Soft wood, such as California redwood, cedar, white pine, and poplar is preferred for nest building.

Control: Infested lumber can be removed and replaced with preservative treated wood. Painting or varnishing wood discourages the bees from boring into it. Untreated wood may be protected temporarily by applying a wettable powder or microencapsualted formulation. These products may have to be reapplied if they are exposed to heavy rain. Dust, wettable powder and aerosol formulations should be applied directly into the galleries. Tunnels shouldbe left undisturbed for several days after treatment, then the holes sealed with wood doweling or putty.

Carpenter Bee

HORNTAIL WASPS

Order/Family:
Hymenoptera/Siricidae

Scientific Name:
Tremex columba (Linnaeus)
Sirex cyaneus Fabricius

Description: Horntails, or wood wasps, are large wasp-like insects that are closely related to bees and wasps. They are 1/2- to 1 5/8-inch long and a deep metallic blue or black color, sometimes with paler markings. The abdomen of the wood wasps are broadly joined to the thorax (segments with legs attached), unlike the true wasps with abdomens joined to the thorax by a narrow junction called a petiole. Females have a long projection from the end of the abdomen that gives these insects the name "horntail." Larvae are found in wood and look much like large wood boring beetle larvae. Mature larvae are 1 3/4-inch long, white, cylindrical, slightly "S" shaped and easily identified by a sharp pointed spine found on the upper side of the last body segment.

Biology: Every ten days, the female wood wasp inserts one to seven eggs into the wood of dead or dying trees using the long egg laying tube (ovipositor) found at the end of her abdomen. The larvae feed on fungus, not the wood, creating cylindrical tunnels throughout its length and producing a sawdust-like frass. They molt three to four times over a two to three year period before pupating near the surface of the wood. The adults emerge five to six weeks later, chew out of the pupal chamber and mate. The life cycle takes two to five years to complete.

Habits: Wood wasps are of great concern when they emerge from lumber used in the construction of homes or other buildings leaving a round hole 1/8- to 1/4-inch in diameter in the wood. The adults deface finished surfaces by boring out through paint, hardwood, floors, linoleum, carpeting, wallboard, plaster, and non-ceramic tile. They cannot reinfest homes but can be discomforting to the building-occupants as they continue to emerge for up to three years after construction is complete.

Control: There is no reason to try to control wood wasp infestations in structures because they do not reinfest. Adults can be vacuumed or killed with an aerosol application as they emerge. Larvae can be controlled in lumber by proper kiln drying or vacuum fumigation before the wood is used in construction. Customers who have wood wasps emerging in their homes should be made aware that normal pest control measures do not prevent wood wasps from continuing to emerge from infested wood. They should be informed that the wasps are harmless and do not reinfest the home. If the customer is overly concerned about further damage and insists that wood wasps be eliminated immediately, the only practical treatment is fumigation. Firewood should be stored outside.

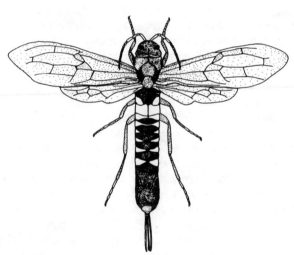

Horntail Wasp

AGGRESSIVE HOUSE SPIDER

Order/Family:	Scientific Name:
Araneae/Theridiidae	*Achaearanea tepidariorum* (C.L. Koch)

Description: The body of the female house spider is about 3/16- to 5/16-inch long and yellow- brown with a dirty-white abdomen with "army sergeant" stripes on the back. Males are much smaller than the females, 1/8- to 3/16-inch long and have a longer, narrower abdomen.

Biology: House spiders lay their eggs in brownish silken sacs which have a tough papery cover. A female produces up to 17 sacs during her lifetime, each containing approximately 250 eggs. The young spiderlings remain in the sac until the second molt. They are cannibalistic and often eat one another. They live in the vicinity of the nest until after the second molt when they produce long threads of silk that help them float away much as kites float. Female spiderlings undergo seven molts before maturing. Adults typically live for a year or more.

Habits: House spiders randomly select web sites, and if the web fails to capture prey, it is abandoned and another is built. They survive better in areas with high humidity, such as garages, sheds, barns, warehouses, etc. The lower humidity in modern structures is not conducive to their survival. However, in structures with higher humidity, webs are constructed in upper corners, under furniture, around window and door frames, basements, garages, and crawl spaces. Outdoors, webs are built around window and door frames, near lights, and under eaves. House spiders feed on a wide variety of insects but especially flies.

Control: Areas conducive to spider activity, e.g., areas that attract insect prey, should be identified. Clutter and debris inside structures and scrap lumber, woodpiles, rocks and other protective outdoor materials should be removed. A vacuum cleaner should be used to remove spiders, webs, and egg sacs and the bag should then be sealed immediately and discarded. Using this technique to control house spiders is relatively easy since they build webs that are more exposed than those of black widows or brown recluse spiders. Outdoor lighting that attracts insects to the structure should be changed.

Pesticides can be applied as residual sprays or dusts with special emphasis on application into potential or known harborage areas. Nonresidual aerosols, mists, and ULV's that contain a pyrethroid insecticide can be used to kill exposed spiders. Dusting spider webs and leaving them undisturbed for several days is also a successful strategy. Pesticides are more effective in eliminating the spiders' food, i.e., insects, than killing the spiders.

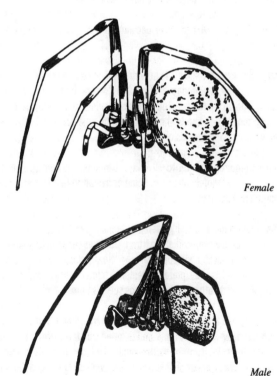

Female

Male

Aggressive House Spider

BLACK WIDOW SPIDERS

Order/Family:
Araneae/Theridiidae

Scientific Name:
Latrodectus spp.

Description: The body of the female black widow spider is about 1/2-inch long, glossy black with a nearly globe-like abdomen. The abdomen has two triangular red spots on its underside arranged in such a way that the spots look like an hourglass. Males are much smaller than the females, 1/4-inch long with a longer, narrower abdomen and somewhat longer legs. Spiderlings are mostly orange and white but become increasingly more black as they mature.

Biology: Black widow spiders lay their eggs in silken sacs which they protect in their nests. A female produces from six to 21 sacs during her lifetime, each containing 185 to 464 eggs. The young spiderlings remain in the case until the second molt. They live in the vicinity of the nest for two to three weeks before producing long threads of silk that help them float away, much as kites float. Female spiderlings undergo from four to nine molts before maturing; this process requires from 54 to 107 days. Development time (i.e., from egg to adult) is approximately one year. Females live up to three years and males approximately 180 days.

Habits: Black widows are shy, preferring to build their webs in dry, protected locations where their prey is likely to travel. Outdoors they can be found among rocks and wood piles, under decks, in hollow stumps, rodent burrows, beneath benches, etc. They prefer basements, crawl spaces, and garages in structures as well as other protected areas such as barns, sheds, meter boxes, brick veneer, pump houses, etc. The webs, which are irregular in shape and approximately one foot in diameter, are used to trap their insect prey which is then paralyzed by their venom. Females often eat the males after mating, thus, giving them their rather morbid name. Females produce a neurotoxin and bite if disturbed or handled roughly. Each year several deaths are attributed to the bite of black widow spiders as a result of anaphylactic reactions, however in most cases, the bite is no worse than a wasp sting.

Control: Areas conducive to spider activity, e.g., dark protected areas that attract insect prey, should be identified. Clutter and debris inside structures and scrap lumber, woodpiles, rocks and other protective outdoor materials should be removed. A vacuum cleaner should be used in order to remove spiders, webs, and egg sacs; the bag should then be sealed immediately and discarded. Outdoor lighting that attracts insects to the structure should be changed.

Pesticides can be applied as residual sprays or dusts with special emphasis on application into potential or known harborage areas. Nonresidual aerosols, mists, and ULV's that contain a pyrethroid insecticide can be used to kill exposed spiders. Dusting spider webs and leaving them undisturbed for several days is also a successful strategy. Pesticides are more effective in eliminating the spiders' food, i.e., insects, than killing the spiders.

Black Widow Spider

BROWN RECLUSE SPIDERS

Order/Family:
Araneae/Loxoscelidae

Scientific Name:
Loxosceles spp.

Description: Brown recluse spiders are light brown or flesh colored to dark brown. They measure 1/4- to 1/2-inch long and have long legs that appear bare. Six eyes are arranged in a semicircle on the top front of the first body segment. The best identifying characteristic is the violin-shaped dark mark that begins right behind the eyes; thus, they are called "fiddle backs" or "violin spiders."

Biology: The female produces from one to five egg sacs in her lifetime, each containing 40 to 50 eggs. The young emerge after about 30 days. Development from egg to adult takes an average of 336 days. The young undergo eight molts. Females live an average of 628 days.

Habits: The brown recluse spider constructs a nondescript, irregular web that is used almost exclusively as a retreat. The nest typically is built in hidden, secluded locations, e.g., among old papers, in seldom-used clothes or shoes, and in attics. Outdoors the brown recluse spider often is found under rocks and in barns and sheds. While hunting, it usually retreats to the nest whenever it is disturbed.

Brown recluse spiders inflict a painful bite that can develop into an ugly, slow-healing ulcer. They seldom bite unless handled or disturbed in their nests. Individuals should be especially cautious of being bitten when donning old, seldom-used clothing.

Control: Areas conducive to spider activity, e.g., dark protected areas which attract insect prey, should be identified. Clutter and debris inside structures and scrap lumber, woodpiles, rocks and other protective outdoor materials should be removed. A vacuum cleaner should be used to remove spiders, webs, and egg sacs, the bag immediately sealed, and discarded. Outdoor lighting that attracts insects to the structure should be changed.

Pesticides can be applied as residual sprays or dusts with special emphasis on application into harborage areas. Nonresidual aerosols, mists, and ULV's which contain a pyrethroid insecticide can be used to kill exposed spiders. Dusting spider webs and leaving them undisturbed for several days is also a successful strategy. Pesticides are more effective in eliminating the spiders' food, i.e., insects, than killing the spiders.

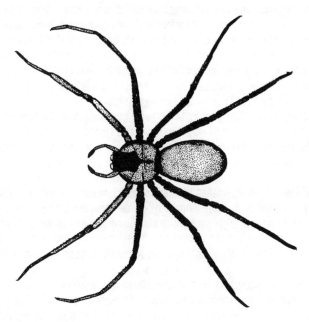

Brown Recluse Spider

YELLOW SAC SPIDERS

Order/Family:
Araneae/Clubionidae

Scientific Name:
Chiracanthium spp.
Trachelas spp.

Description: The bodies of female sac spiders are about 3/16- to 3/8-inch long and either pale yellow to pale green or orange brown to red. Males are smaller than the females, 1/8- to 5/16-inch long.

Biology: Sac spiders lay their eggs in a loose mass covered with a thin white silk sac. Eggs often are laid in indoor retreats and are guarded by the female; however, eggs can be found outdoors in rolled leaves. Females produce several egg masses during their lifetimes, each containing 30 to 48 eggs.

Habits: Sac spiders are of medical importance because, as a group, they are more responsible than any other species for spider bites in the United States, but bites often go unreported because the spider goes unseen or the bite is not felt. Most bites occur when the spider crawls into clothing and bites when pressed close to the skin. In most cases, the bite produces no more than localized redness, slight swelling, and a burning sensation at the site of the bite.

Sac spiders are commonly found indoors. Their numbers increase significantly in the fall when the weather turns cool and their food supply disappears. They enter structures through faulty screens, cracks around doors and windows, and gaps around pipes, wires, and vents in exterior walls. If food is available, they enter and remain in wall voids and crawlspaces. They build silk retreats in upper corners and the joints between walls and ceilings indoors and remain in them during the day. Outdoors they are found among rocks and wood piles, under decks and benches, around window and door frames, roof eaves, soffits, behind shutters, and in protected areas, e.g., barns, outbuildings, sheds, meter boxes, brick veneer, pump houses, etc. They are also found on tall grass, weeds, and leaf litter.

Control: Areas conducive to spider activity, e.g., dark protected areas that attract insect prey, should be identified. Clutter and debris inside structures and scrap lumber, woodpiles, rocks, landscape timbers, and other protective outdoor materials should be removed. Both indoors and out, a vacuum cleaner with a crack and crevice attachment should be used in order to remove spiders, sacs, and egg sacs; then the bag should be sealed and discarded immediately. Particular attention should be paid to upper corners where all sacs should be removed. Trees and bushes should be trimmed back so that they do not contact the structure, and the grass should be mowed.

Pesticides can be applied as residual sprays or dusts with special emphasis on application into harborage areas. Nonresidual aerosols, mists, and ULV's that contain a pyrethroid insecticide can be used to kill exposed spiders. Dusting hollow block voids along the sill plate might be useful in reducing indoor populations. Pesticides are more effective in eliminating the spiders' food, i.e., insects, than killing the spiders.

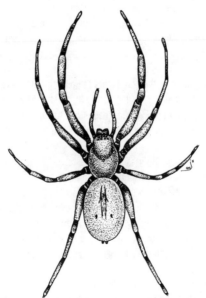

Yellow Sac Spider

HONEY BEE

Order/Family:
Hymenoptera/Apidae

Scientific Name:
Apis mellifera Linnaeus

Description: Honey bees have three castes in their colonies: workers, queens, and drones. Workers are 1/2- to 5/8-inch long, have well developed heads with two short antennae, and large eyes. These insects have a fuzzy yellow-brown to black appearance because they are covered with fine hairs. The abdomen has a striped appearance. They have two pairs of wings, the hind pair shorter than the front pair. The workers have a barbed stinger at the end of their abdomen that is used against anything that threatens the colony.

The back pair of legs is modified for the collection and transport of pollen. Honey bees have a tongue-like mouthpart which allows them to collect nectar in flowers. Queen bees are the largest member of the colony, measuring from 5/8- to 3/4-inch long and except for their size look like the workers. Drones are about 5/8-inch long and much stouter and darker than workers or the queen.

Biology: The queen is the only fertile female; she produces all the eggs for the colony. The queen is capable of producing 1,000-2,000 eggs per day. Drones serve only to fertilize the queens and are driven off by the workers after they have served that purpose. Workers live five to seven weeks during the summer.

Habits: Honey bees are social insects that live in the colony or hive with as many as 20,000-80,000 workers. Workers collect nectar and pollen from plants, inadvertently pollinating flowers and allowing plants to produce fruit. They also produce honey and fashion honey comb from wax that they secrete. The queen and all the bee larvae are fed and cared for by young workers. Older workers gather the pollen and nectar for the colony. The entire population overwinters.

Honey bees are not naturally aggressive; however, if the colony is threatened they will sting. Honey bees swarm when the queen begins to fail or the colony is too large. Swarms often are seen on a tree branch, and when this occurs, the bees are not aggressive. The swarm lasts 24-48 hours and then moves to a sheltered environment, e.g., hollow tree, bee hive, hollow wall, attic, etc.

Control: If a honey bee colony is established within the wall of a structure, there are two control options: the living colony can be removed or killed. If the decision is to remove a living colony, it is best to work with an experienced beekeeper who will help remove the bees safely. Swarms are easily removed by an experienced bee keeper. Removal of a colony within a structure may require several weeks and frequently the building occupants are not willing to wait that long for this solution to the problem.

Because native honey bee colonies are rapidly declining, killing the colony should be avoided, if possible. If necessary, the colony can be killed by applying residual insecticide dust or wettable powder formulations around the entrance to the colony. The entrance of the nest should be left open so that the bees are able to leave the nest and die. Once the colony has been killed, the wall should be opened to remove the honey comb and developing larvae which prevents the development of secondary pest problems, i.e., wax moths, dermestids, and other scavengers.

Honey Bee

SOCIAL WASPS

Scientific Name:
Vespula spp. — Yellow Jackets
Vespula maculata (Linnaeus) — Baldfaced Hornet
Vespa crabro (Christ) — European Hornet
or Giant Hornet

Order/Family:
Hymenoptera/Vespidae

Polistes spp. — Paper Wasps

Description: Social wasps have the typical "wasp" body type: a very distinct head with chewing mouthparts, short-elbowed antennae, and large compound eyes. The thorax and abdomen are brightly marked with yellow, red, or brown on a black background. The wasps have four clear or smoky brown wings. They have a short, narrow attachment between the thorax and the abdomen. The abdomen is spindle-shaped and tipped with a long stinger.

Yellow jackets are usually marked with bright yellow and black patterns, appear hairless, and are about 3/8- to 5/8-inch long. The baldfaced hornet is similar in appearance except that it is white and black and 5/8- to 3/4-inch long.

The giant European hornet is 3/4- to 1 3/8-inch long and brown and yellow. Paper wasps can be distinguished from yellow jackets and hornets because their abdomen is tapered at the tip and at the point where it joins the thorax. They are sometimes marked with yellow, brown, or red patterns on black and are 5/8- to 3/4-inch long.

Biology: Social wasps have large nests containing three types of individuals, or castes: queens, workers, and males. The males and queens are produced in the colony in late summer. They mate, and the fertilized queen overwinters in a protected site. In spring, she seeks an appropriate nesting site in which she builds a paper nest using chewed up wood fibers. Eggs are laid in the cells of the nest, and the young larvae are fed bits of chewed up meat or insect parts by the queen and later by the workers.

Habits: Yellow jackets and hornets build their flat paper nests in stacks which are surrounded by a paper envelope. Yellow jackets usually build their nests below ground and in other protected locations. Baldfaced hornets prefer to build their nests in trees and on the sides of buildings. European hornets build their nests inside hollow trees, wall voids, and underground. Paper wasps build open flat nests, without a paper envelope, usually found under the eaves of a house and in other protected locations. Social wasps use their nests only one season.

Unlike bees, these wasps aggressively defend their nests and can inflict multiple stings. They produce very large colonies with some yellow jacket nests containing as many as 30,000 individuals. These insects are considered to be beneficial because they feed their young a wide variety of insects. They become a nuisance, however, when they build nests in or near structures; scavenge for food in recreational areas and in other places frequented by humans; and seek overwintering sites in structures.

Control: When dealing with social wasps, protective equipment including a bee hat, long-sleeved shirt, coveralls, eye wear, and gloves should be worn. The nest should be located by examining all protected areas in the vicinity of wasp activity. Simply removing the nest will not resolve the problem because surviving wasps will construct a new one.

Dust, aerosol, and liquid formulations are effective against social wasps. Aerosols in the form of "wasp freezes" are formulated for instant knock-down and for applications from 10 to 20 feet away. Dusts are very effective if injected into the opening of the nest. A light application is enough to eventually kill all the wasps as they pass through the opening.

The best strategy is to treat the nest at night when all the workers and queen are present. This tactic maximizes the effect of the pesticide application by killing most, if not all, of the wasps. If treatment is made at night, shining a light directly on the nest should be avoided or a red filter should be used on the flashlight.

Daytime treatments are successful when the nest is treated and if the wasps present on the nest are killed, the nest is removed, and the nest attachment area treated. Returning workers searching for the nest contact the residual pesticide and die.

The job should be completed by removing the nest, particularly if it is inside an attic, wall void, etc. This service prevents secondary infestations by dermestids or other pests.

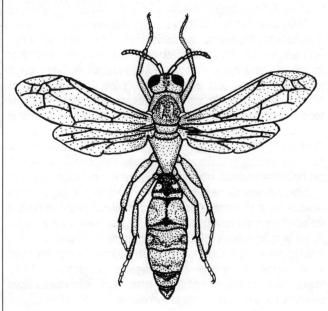

Social Wasp

SOLITARY WASPS

Order/Family:
Hymenoptera/Pompilidae — Spider Wasps
Hymenoptera/Sphecidae — Mud Daubers
and Digger Wasps
Hymenoptera/Vespidae — Potter Wasps

Scientific Names:
Various

Description: Solitary wasps vary in size from 1/4- to 2-inches in length. Their color also varies from dull black or brown to brilliant red, yellow or blue. Many have a metallic sheen to the body or wing. They have a typical wasp body type, frequently with a long slender petiole (i.e., connection) between the abdomen and the thorax. Although they have stingers, they usually are not aggressive, stinging only when handled.

Biology: The adults emerge in the spring, mate, and begin construction of their nests which may contain one or more cells. A single egg is laid in each cell and provisioned with captured prey. When the egg hatches, the larva feeds on the captured prey. Once development, which requires approximately three weeks, is complete, the larva spins a silken cocoon and overwinters until the following spring when it pupates.

Habits: Solitary wasps are predators that capture and sting insects and spiders to provision their nests. The type of nest constructed depends on the species. Potter wasps build tiny jug-shaped mud nests attached to twigs or structures and provisions the nest with caterpillars.

Spider wasps burrow in the ground and provisions the nest with spiders which it hides until the nest is completed. Mud daubers build a variety of mud nests, including organ-pipe nests and the globular nest built by the black and yellow mud dauber. Digger wasps burrow in the soil. The cicada killer is the most spectacular of the group because of its large size and the the the cicadas it captures as prey. Solitary wasps generally are considered to be beneficial insects. However, many customers become alarmed by their presence when they build nests on buildings and burrow in lawns, flowerbeds, and gardens.

Control: The beneficial nature of these wasps should be explained to the customer and, if possible, killing them should be avoided. However, if the customer demands elimination, the following procedures are recommended: Mud dauber and potter wasps can be eliminated easily by removing the nests and killing the adults with an aerosol product. Digger wasps, cicada killers, and burrowing spider wasps can be controlled by applying residual dusts and wettable powders into the burrows.

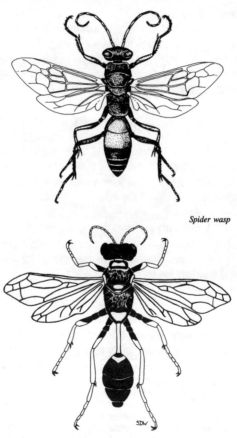

Spider wasp

Solitary Wasps

Mud dauber

SCORPIONS

Order/Family:
Scorpiones/Various

Scientific Name:
Various

Description: Scorpions are close relatives of spiders having broad, flat bodies with eight legs. Adults are 2 to 4 inches long and readily identified by their two pincer-like pedipalps at the front of their bodies and the five-segmented stinger-tipped tail at the back. Depending on the species, scorpions range in color from the mustard yellow of the deadly sculptured scorpion to the black of the black scorpion; typically they are mottled to a striped brown tan in color.

Biology: A female scorpion does not lay eggs, but produces an average of 35 live nymphs per brood which she carries on her back for from seven to 30 days. The nymphs undergo an average of six molts over a period of several months to four years before reaching maturity. Adults live for one to six years. Scorpions are predators, feeding primarily on insects and spiders, and are able to survive up to six months without feeding.

Habits: Though scorpions are most commonly found in the South, especially in desert areas of the Southwest; however their northern occurrence extends along a line from Baltimore to St. Louis to San Francisco. Scorpions are poisonous; the poison glands are in the bulbous last segment of the tail. Most species are not dangerous but inflict a sting comparable to that of a wasp. However, the deadly sculptured scorpion common in Arizona, has been responsible for many deaths.

Scorpions are active at night, feeding on spiders and insects. During the day they hide under stones and tree bark, in rock and wood piles, and in masonry cracks. They enter structures seeking water and shelter and are frequently found in bathrooms, crawlspaces, attics, dry stone walls, foundations, and in clothes and shoes left on the floor.

Control: Scorpions that cause a nuisance outdoors can be controlled by removing harborage areas, e.g., trash piles, stones, boards, firewood stored off the ground, landscape timbers, debris, etc. Entry into structures can be prevented by sealing and caulking gaps around siding, windows, doors, pipes, and wires.

Microencapsulated and wettable powder products, which also eliminate insects and other arthropods eaten by scorpions, are the most effective formulations that can be used in the outdoor habitats preferred by scorpions. These products should be applied in a three- to ten-foot band around the perimeter of the structure, into harborage sites, and/or around potential entry points.

Indoors scorpions can be controlled by sealing their harborage areas wherever possible. A vacuum cleaner should be used to remove exposed scorpions. Dust applications can be introduced into wall voids, voids between floors, attics, and other potential hiding places. Keeping scorpions out of living areas and eliminating insects and other arthropods eaten by scorpions is the key to controlling this pest around structures.

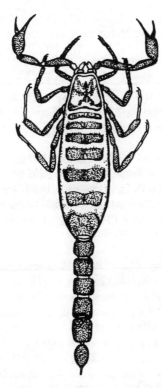

Scorpion

BATS

Scientific Name:
Eptesicus fuscus — Big Brown Bat

Order/Family:
Chiroptera/Various

Myotis lucifugus — Little Brown Bat
Tadarida brasiliensis — Mexican/Brazilian Free-tailed Bat

Identification: These bat species typically are found inside structures. Depending on the species, the adults are 2 1/4- to 7 1/2-inches long with a wingspread of six- to 15-inches. They are tan to black, have large ears, with hair-covered bodies.

Biology: Nursery colonies typically are formed in locations other than the overwintering site. Depending on the species, one or two babies are born from May through July. The young hang in the roost or are carried by the female. They are weaned and able to fly within a month and leave the roost by the end of August. The average life span of most bats is four to 10 years; some, however, live as long as 19 to 38 years.

Habits: Big brown bats are found throughout North America except southern Florida. Little brown bats are found in the same areas except for Texas and southern California. Mexican free-tailed bats are found throughout the southern United States.

Despite the benefit that bats consume thousands of insects per night, humans fear them due to superstition as well as their potential health threat to humans. Rabies and histoplasmosis occasionally are of concern when dealing with bat infestations in structures. In most colonies, only a small percentage of the bat population is infected with rabies, and because bats avoid contact with humans and domestic animals, the incidence of human exposure to bat rabies is low. As bat droppings accumulate in the roost, the risk of histoplasmosis increases.

All three species leave their roosts at dusk in order to feed and continue to do so until just before dawn. Initially they visit ponds, streams, and other water sources in order to drink and then begin their nightly feeding frenzy. The roosting habits of the bats vary according to species. When nursery colonies of large brown bats are first formed, the males roost separately; later in the summer, however, all roost together. In early fall, small brown bats migrate to roosts within caves and mines and return to structures in early spring. Mexican free-tailed bats, which form large colonies, vary in their habits, some migrating to Mexico for the winter, others roosting indefinitely within structures.

Control: An inspection to determine where the bats are entering the structure should be conducted approximately one half hour before dusk until one hour later. The inspector should look for gaps around the fascia boards, vents, soffits, chimneys, roof lines and other areas. Attic areas should be inspected to determine the size of the colony, other points of entry, the amount of accumulated droppings, and the need for dropping removal and disinfection.

Exclusion, the best long-term solution for bat infestations, should be used only from September through April after the bats have migrated or when dealing with non-migratory species in order to avoid trapping newborn bats within the structure. All entry points greater than 3/8-inch wide should be sealed. The most effective procedure is to seal all but one or two openings, thereby allowing three to four days for the bats to adjust, and then to seal the remaining holes one evening after they emerge from the structure. One-way flaps can be installed to permit the bats to exit but not reenter the structure. Entry holes should be sealed with 1/4-inch hardware cloth, sheet metal, or metal wool. In situations where there are too many holes to seal, e.g., Spanish barrel tile roofs, these areas can be covered with small mesh bird netting.

Using moth balls and light in order to repel bats is of limited value because the bats typically relocate within the present structure. Ultrasonic devices are ineffective in repelling bats. If a stray bat enters the interior of a structure, it usually can be encouraged to leave through an open door or window. It also could be captured, e.g., under a coffee can using a piece of cardboard as a lid, and releasing the bat unharmed outdoors. Under no circumstances should the bat be handled. A residual insecticide and/or ULV application should be made to the roost area in order to control ectoparasites, e.g., bat bugs and mites. Excessive dropping accumulations should be removed and the area disinfected.

Bat

HOUSE SPARROW

Order/Family:
Passeriformes/Passeridae

Scientific Name:
Passer domesticus

Description: The house, or English, sparrow is one of two species that occurs in North America. It is not actually a sparrow but a type of finch. House sparrows are approximately six inches long. Their color varies with the sex, males having a black patch under the beak and the cheeks, rump, and top of the head grey-white. In the winter the black area is hidden by the grey feather tips. Females and young sparrows are dull brown with a dirty white breast and brow.

Biology: The eggs range in color from white to light green to light blue and have numerous dark spots. They average five to six eggs per clutch and require ten to fourteen days to incubate. Two to three broods per year or up to 21 young per year are produced. After hatching the almost featherless young birds are totally dependent on the parent birds for food and warmth.

Habits: Sparrows prefer to nest in protected man-made or natural areas. Building ledges, openings in structures, gutters, signs, light fixtures, birdhouses, under the eaves of a house, bridges, electrical power lines and transformers are areas which often are used to construct nests. Sparrows also displace other birds (e.g. wrens, robins, and purple martins) from their nests; destroy their eggs and use their nests for their own brood. Occasionally, nests are constructed in trees. Both sexes construct a rather large and flimsy nest from straw, grass, feathers, strips of paper, string and other debris. House sparrows frequently nest in kitchen, bathroom, oven, and dryer vents.

Sparrows tend to be very territorial, as individuals and as flocks and restrict their nesting and feeding to specific locations. They congregate in urban areas in the winter and disperse to rural areas in the spring. Flocks of juvenile birds and non-breeding adults sometimes travel four to five miles from nest sites to feeding areas.

They foul structures with their droppings, particularly those areas used for roosting and loafing sites. Sparrows also enter food plants, warehouses, department stores and malls where they often contaminate food products or other merchandise. Their droppings can contain variety of disease causing bacteria, fungi, nematodes, etc. Sparrows are considered to be one of the major reservoirs of St. Louis encephalitis. Numerous blood-feeding parasitic mites associated with sparrows also bite humans.

Control: The most important step in bird control work which is the survey which is used to determine species, population size, location of the problem, habitat, resources, activity patterns, time constraints, public relations issues, equipment, etc. The site should be visited on several days and at different times to better analyze activity patterns.

Both nonchemical and chemical means of eliminating bird problems exist. Sanitation, which removes the food, water, and protection the birds seek, is an important first step. Exclusion involves structural modifications which deter roosting and nesting. Options include plastic and metal spikes, plastic netting, pin and monofilament line/wire, and the use of repellent gels and pastes. Sound devices might be effective but are considered unsuitable for residential environments. Trapping often is slow, labor-intensive, and effective only if the birds are destroyed instead of released. Shooting, a final alternative, for obvious reasons has limited application.

There is one pesticide available for bird control. It is a chemical frightening agent formulated as a bait which, when eaten, causes affected birds to emit distress calls, thus, repelling the remainder of the flock. It may be lethal to birds that consume too much bait. Use of the product requires prebaiting and frequent service.

House Sparrow

PIGEON

Order/Family:
Columbiformes/Columbidae

Scientific Name:
Columba livia Gmelin

Description: The rock dove, more commonly referred to as a pigeon, is one of eleven species that breed in North America. This bird is approximately thirteen inches long, with color varying from white to black. However, the rock dove is characterized by its dark grey head with an iridescent sheen, light grey back, and wings with two dark bands. The rock dove has a stocky body with short legs and neck and a small head. During flight, the tip of the tail is usually square and black in color.

Biology: The white eggs are laid one-to-two at a time and require approximately 18 days to incubate. Two to five broods are produced per year or up to ten young per year. The eggs are incubated by both parents.

After hatching, the almost featherless young birds are totally dependent on the parent birds for warmth and food. For the first five days, the newly hatched birds are fed a milky substance, "pigeon milk," which is produced in the crop of the parent birds. During the next five days, more water and grain is incorporated into the "milk," and, finally, they are fed only grain and water. The young pigeons leave the nest approximately one month after they hatch. Pigeons live from four to twelve years.

Habits: Pigeons have become the most serious bird pest associated with buildings. They nest in a variety of protected locations, such as the underside of bridges, building ledges, rafters in barns and other open buildings, roofs, air conditioners, signs, etc. The loosely-constructed nests typically consist of sticks, stems, leaves, and other debris. Nests that are reused often become solid with the accumulation of droppings and debris. Pigeons nest during all seasons when conditions permit.

In rural settings, pigeons typically feed on seeds, grain and fruit. They find areas with spilled silage, such as grain elevators, railroad yards, and mills, all of which are very attractive feeding sites. In urban areas, pigeons feed on handouts, garbage, vegetable matter, and insects.

In contrast to many other bird species, pigeons prefer flat and smooth surfaces, such as roof tops, for resting and feeding. This affords them a quick get-away if they are threatened. Pigeons gather in flocks which use the same roosting and feeding areas. Feeding usually occurs no more than a few miles from the roosting site.

Their very acidic droppings can cause significant damage to equipment, painted building surfaces, marble, limestone, etc. Droppings, which also contaminate unprocessed grain and processed food and can contain a variety of disease-causing bacteria, fungi, nematodes, etc. Pigeons serve as reservoirs for several viral encephalitic diseases. Many of the parasitic mites associated with pigeons also bite humans.

Control: The most important step in bird control work is the survey which is used to determine species, population size, location of the problem, habitat, resources, activity patterns, time constraints, public relations issues, equipment, etc. The site should be visited on several days and at different times to better analyze activity patterns.

Both nonchemical and chemical means of eliminating bird problems exist. Sanitation, i.e., removing the food, water, and protection that the birds seek, is an important first step. Exclusion involves structural modifications which deter roosting and nesting. Options include plastic and metal spikes, plastic netting, pin and monofilament line/wire, and the use of repellent gels and pastes. Sound devices might be effective but are considered unsuitable for residential environments. Trapping is often slow, labor-intensive, and effective only if the birds are destroyed instead of released. Shooting, the final alternative, for obvious reasons, has limited application.

There is one pesticide available for bird control. It is a chemical frightening agent formulated as a bait which, when eaten, causes affected birds to emit distress calls, thus, repelling the remainder of the flock. It may be lethal to birds that consume too much bait. Use of the product requires prebaiting and frequent service.

STARLING

Order/Family:
Passeriformes/Sturnidae

Scientific Name:
Sturnus vulgaris Linnaeus

Description: The European starling is one of two species that were introduced into North America. Starlings are seven to eight inches long. Their color varies with the season — they are purplish-black with an iridescent sheen in the summer; in the winter the tips of the feathers are marked with white and gold giving them a speckled appearance. The long pointy bills are bright yellow in the spring and summer but turn dark in the winter. Their stocky bodies and very short tails make them appear tailless.

Biology: The eggs vary in color, from white to light blue; some have dark spots. Both parents are involved in building the nest, incubating the eggs, and caring for the young. The eggs are laid two to eight at a time and require 12-14 days to incubate. They average two broods per year, producing up to 16 young per year. After hatching the almost-featherless young are totally dependent on the parent birds for food and water.

Young starlings leave the nest approximately three weeks after they hatch. Unmated males flock and move from roosts to feeding sites together. As the first brood matures, they join this flock. Thus, as late summer approaches, the flocks increase significantly as the final brood and mating pairs join them.

Habits: Starlings typically select nesting sites that are in the shadows of brighter light. In urban areas they tend to roost in building cavities, often 20 to 70 feet above the average street light height. In suburban and rural settings, they often nest in tree holes, birdhouses with holes larger than 1 1/2-inches in diameter, and other protected areas two to 60 feet above the ground. Nests are constructed of twigs, grass, and other debris, then lined with feathers and other soft materials. In urban areas, European starlings are increasingly found nesting in kitchen, bathroom, oven, and dryer vents.

At dawn, starlings travel as far as 70 miles from the roosting site to a feeding site. When they return to the roosting area at dusk, they first perch on telephone wires, bridges, buildings, and trees until after sunset at which time they fly around the roosting site, perhaps several times, before settling in for the evening. Some starlings migrate as cold weather approaches. Birds that do not migrate usually roost in protected areas, such as buildings in urban areas.

Starlings feed on the ground and away from their roosting sites. During spring and early summer the nesting birds eat insects and occasionally soft fruit. During late summer, fall, and winter, their diet preference shifts to grains, seeds, and fruits. They can consume up to one ounce of grain per day.

Control: The most important step in bird control work is the survey which is used to determine species, population size, location of the problem, habitat, resources, activity patterns, time constraints, public relations issues, equipment, etc. The site should be visited on several days and at different times to better analyze activity patterns.

Both nonchemical and chemical means of eliminating bird problems exist. Sanitation, which removes the food, water, and protection that the birds seek, is an important first step. Exclusion involves structural modifications which deter roosting and nesting. Options include plastic and metal spikes, plastic netting, pin and monofilament line/wire, and the use of repellent gels and pastes. Sound devices might be effective but are considered to be unsuitable for residential environments.

There is one pesticide available for bird control. It is a chemical frightening agent formulated as a bait which, when eaten, causes affected birds to emit distress calls, thus, repelling the remainder of the flock. It may be lethal to birds that consume too much bait. Use of the product requires prebaiting and frequent service.

DEER MOUSE

Order/Family:
Rodentia/Muridae

Scientific Name:
Peromyscus maniculatus (Wagner)

Identification: Although the genus of the deer mouse is rather large, each species has its own common name. Members of the genus are referred to as either deer mice or white-footed mice. The most common and widely-distributed species is *P. maniculatus*, the true deer mouse.

The deer mouse is bi-colored — the upper portion of the body and tail is medium- to dark-brown, and the underside of the tail, feet and stomach areas are white. The body is 2 3/4- to 4-inches long and the tail two to five inches long. The eyes, ears and body of a deer mouse are slightly larger than those of the house mouse.

Biology: Deer mice produce from three to four litters, each of which contains three to five young. Thus, populations build up rapidly. They typically produce their largest litters in the spring, depending on climatic conditions and begin to breed at five to six weeks of age. The life span is from two to 24 months.

Habits: Deer mice are found throughout the United States. This species gained national notoriety last summer when public health officials determined that the deer mouse is the principle rodent species associated with the transmission of the hantavirus. Hantavirus is transmitted through the inhalation of particulate matter which is contaminated by the droppings and urine of infected mice. Disease mortality in humans is approximately 60%.

Deer mice are active year round. Their range of activity is one-half to three acres. They often construct nests in hollow logs and tree stumps, under logs and stones, and occasionally in bird nests and shallow burrows. Deer mice rarely are a major problem in residential areas; however, housing in rural and agricultural areas may have more of a problem.

In the fall and winter, deer mice enter houses, garages and outbuildings, and occasionally campers and other infrequently-used vehicles. Once inside these areas, they can cause significant damage to furnishings and stored materials as they search for food and construct their nests. Their typical diet consists of nuts, seeds, berries and insects. They often store food in their nests for the winter months. They are nocturnal and are rarely seen in their outdoor habitat.

Control: The best solution for problems with deer mice is exclusion. Entry holes should be sealed with 1/4-inch hardware cloth, sheet metal, or metal wool and particular attention paid to any hole which approximates the diameter of a pencil. In some situations, perimeter trapping might prevent a population from becoming established indoors. Pet foods and other food products should not be stored in accessible areas, such as garages, and water sources which also attract mice should be eliminated.

Once deer mice enter structures, they can be effectively controlled by using baited and unbaited snap traps and glueboards, anticoagulant rodenticides, and tracking powders. Their inquisitive nature makes them easy to trap.

Due to increased concern about the hantavirus, deer mouse populations in structures should be eliminated as quickly as possible. Dead rodents and their nests and droppings should be removed immediately.

HOUSE MOUSE

Order/Family:
Rodentia/Muridae

Scientific Name:
Mus musculus Linnaeus

Identification: The house mouse is the most common and economically important commensal (i.e., living in close association with humans) rodent. The house mouse is gray and it weighs one half to one ounce. The body is three to four inches long and the tail three to four inches long. The muzzle is pointed, the ears are large, the eyes and body are small. Typically, the house mouse is slightly smaller than deer mice. Adult droppings are 1/8- to 1/4- inch long and rod-shaped with pointed ends.

Biology: The female house mouse reaches sexual maturity in 35 days and averages eight litters per year, each of which averages six young. Thus, with 30-35 weaned mice per year, populations build up rapidly. They typically produce their largest litters in the spring, depending on climatic conditions and begin to breed at five to six weeks of age. The life span is one year.

Habits: House mice are found throughout the United States. They are good climbers, jump 12 inches high, and can jump down from eight feet. House mice easily squeeze through holes and gaps wider than 1/4-inch. They are very social in their behavior, very inquisitive about things in their environment, and readily explore anything new.

House mice prefer to nest in dark secluded areas where there is little chance of disturbance, and in areas where nesting materials, such as paper, cardboard, attic insulation, cotton, etc., are readily available. Their foraging territories are small usually no more than 20 feet; however, if abundant food is nearby they nest within four to five feet. They nibble on food, preferring items such as seeds and cereals. They feed at dusk and just before dawn.

The major health risks associated with house mice are salmonella contamination and leptospirosis.

Control: During the inspection look for signs of activity, such as droppings and rub marks; however, marks left by house mice are less noticeable than those produced by rats. Entry holes should be sealed with 1/4-inch hardware cloth, sheet metal, or metal wool, payng particular attention to any hole which approximates the diameter of a pencil. Pet foods and other food products should not be stored in accessible areas, such as garages, and water sources which also attract mice, should also be eliminated.

Within structures, house mice can be controlled effectively by using baited and unbaited snap traps and glueboards. Traps can be baited with nesting materials, such as cotton, string, dental floss, etc., or fruit, vegetables, and seeds. Their inquisitive nature makes them easy to trap; thus, periodically moving the traps increases success.

Several anticoagulant rodenticides are available as pellets, packets, and blocks. When baiting indoors, these products should be placed in tamper-resistant bait stations and in areas which are inaccessible to children and pets. When water sources are limited, liquid anticoagulant baits are very effective but must be handled in the same way as dry baits. The attractiveness of liquid baits can be increased by formulating the rodenticide with a low concentration fruit juice. Tracking powders are also available which contain either an acute single-dose toxicant or anticoagulant active ingredients.

NORWAY RAT

Order/Family:
Rodentia/Muridae

Scientific Name:
Rattus norvegicus (Berkenhout)

Description: The Norway rat is the largest of the commensal (i.e., living in close association with humans) rodents. The head and body are seven to ten inches long and the tail is an additional six to eight inches. It has a stocky body and weighs seven to 18 ounces. The fur is coarse, shaggy, and brown with some black hairs. The muzzle is blunt, eyes and ears are small, and the tail, which is bi-colored, is shorter than the head and body combined. Norway rat droppings are up to 3/4-inch long with blunt ends.

Biology: Adults are sexually mature in two to five months. Females produce three to six litters per year, each averaging seven to eight young. Adults live from six to twelve months. They have poor sight but keen senses of smell, taste, hearing, and touch.

Habits: Rats are nocturnal. They are shy about new objects and very cautious when things change in their environment and along their established runs. Outdoors, Norway rats prefer to nest in burrows in the soil, e.g., under sidewalks and concrete pads, stream/river banks, railroad track beds, next to buildings, in low ground cover, etc. The burrows typically have one main entry hole and at least one escape hole. The rats easily enter buildings through 1/2-inch and larger gaps. In buildings they prefer to nest in the lower levels of the building, e.g., crawlspace, basement, loading dock and sewers. They prefer foods such as meat, fish, and cereals and require a separate nonfood water source. Their foraging range is 100 to 150 feet from their nest. Rats are associated with various diseases and occasionally bite. Plague is of little concern because it has not occurred in rats in the United States for many years. However, leptospirosis is vectored by rats, and, thus, is a disease of great concern. This disease is acquired by eating food and drinking water which are contaminated with infected rat urine. Rats also cause significant structural damage and product destruction.

Control: The keys to a successful program of rodent control are identification, sanitation, elimination of harborage, and rodent-proofing. The inspection should identify signs of infestation, e.g., gnaw marks, droppings, tracks, burrows, rub marks (i.e., dark greasy spots left where the rats rub against surfaces), runways, damaged goods, etc. Sanitation consists of removing food, water, and materials which provide harborage. Stored goods should be at least twelve inches off the floor and eighteen inches away from the wall.

Rats can be trapped using glueboards and snap traps placed along walls and near vertical runs where the rats travel. Traps can be baited with fish, meat, and cereal, or they can be left unbaited. Because rats are wary of new things in their environment, it may be helpful to leave the traps unset for awhile.

Exclusion is a critical aspect of rodent control. Since rats can squeeze through a 1/2-inch gap, anything larger must be sealed. Since they can chew holes, sealing smaller holes should be considered. Sheet metal, cement, 1/4-inch hardware cloth, expandable foams, etc. are the materials of choice which can be used for this service. Door sweeps should be installed around gaps on doors, windows, and other openings.

Several anticoagulant rodenticides are available as pellets, packets, and blocks. When baiting indoors, these products should be placed in tamper-resistant bait stations and in areas which are inaccessible to children and pets. Outdoors, pellet baits can be placed deep in the burrow using a long-handled spoon and then the burrow should be closed. If packets and blocks are used in burrows, there is a risk that the rat will push the product outside the burrow. When water sources are limited, liquid anticoagulant baits are very effective but must be handled in the same way as dry baits. Tracking powders are also available which contain either an acute single-dose toxicant or anticoagulant active ingredients and typically are used to dust burrows and runs. Using gas in outdoor burrows is an option but is risky particularly if the burrow extends under or into a structure.

ROOF RAT

Order/Family:	Scientific Name:
Rodentia/Muridae	*Rattus rattus* Linnaeus

Description: The roof rat is a commensal (i.e., living in close association with humans) rodent. The head and body are six to eight inches long and the tail is an additional seven to ten inches. It has a slight body which weighs five to nine ounces. The fur is soft, smooth, and brown in color with some black hairs. The muzzle is pointed, eyes and ears are large, and the scaly tail, which is uniformly dark, is longer than the head and body combined. Roof rat droppings are up to 1/2-inch long and spindle-shaped with pointed ends.

Biology: Adults are sexually mature in two to five months. Females produce four to six litters per year, each averaging six to eight young. Adults live from nine to 12 months. They have poor sight but keen senses of smell, taste, hearing, and touch.

Habits: Rats are nocturnal. They are shy about new objects and very cautious when things change in their environment and along their established runs. Outdoors, roof rats prefer to nest in trees and, occasionally, in burrows and vegetation. The rats easily enter buildings through 1/2-inch and larger gaps. In buildings they prefer to nest in the upper levels of the building, and, occasionally, in basements and sewers. They prefer foods such as fruits, vegetables, and cereals. Their foraging range is 100 to 150 feet from their nest.

Rats are associated with various diseases and occasionally bite. Plague is of little concern because it has not occurred in rats in the United States for many years. However, leptospirosis is vectored by rats, and, thus, is a disease of great concern. This disease is acquired by eating food and drinking water which are contaminated with infected rat urine. Rats also cause significant structural damage and product destruction.

Control: The keys to a successful program of rodent control are identification, sanitation, elimination of harborage, and rodent-proofing. The inspection should identify signs of infestation, e.g., gnaw marks, droppings, tracks, burrows, rub marks (i.e., dark greasy spots left where the rats rub against surfaces), runways, damaged goods, etc. Sanitation consists of removing food, water, and materials which provide harborage. Stored goods should be at least twelve inches off the floor and eighteen inches away from the wall.

Rats can be trapped using glueboards and snap traps placed along walls and near vertical runs where the rats travel. Traps can be baited with fruit, vegetables, and cereal, or they can be left unbaited. Because rats are wary of new things in their environment, it may be helpful to leave the traps unset for awhile.

Exclusion is a critical aspect of rodent control. Since rats can squeeze through an inch gap, anything larger must be sealed. Since they can chew holes, sealing smaller holes should be considered. Sheet metal, cement, 1/4-inch hardware cloth, expandable foams, etc. are the materials of choice which can be used for this service. Door sweeps and other materials should be installed around gaps on doors, windows, and other openings.

Several anticoagulant rodenticides are available as pellets, packets, and blocks. When baiting indoors these products should be placed in tamper-resistant bait stations and in areas which are inaccessible to children and pets. When water sources are limited, liquid anticoagulant baits are very effective but must be handled in the same way as dry baits. Tracking powders are also available which contain either an acute single-dose toxicant or anticoagulant active ingredients.

The PCT Technical Resource Library

PCT Field Guide for the Management Of Structure-Infesting Ants – Second Edition

By Stoy Hedges

Completely revised and updated, the 2nd edition of this popular book is the most comprehensive ant field guide available in the pest control industry. This valuable field guide features the latest industry research on ants and updated profiles of several important ant species not featured in the first edition. The book contains more than 80 photographs and illustrations, as well as individual profiles for commonly encountered one- and two-node ants.

PCT Field Guide for the Management Of Structure-Infesting Flies
Also available in Spanish Language Edition!

By Stoy Hedges

A comprehensive publication featuring in-depth profiles of the three classes of flies frequently encountered by PCOs: small flies, filth flies and nuisance flies. The book includes identification tips, a brief taxonomic key, practical management strategies, case histories and a listing of leading fly control manufacturers. The guide offers a special color photograph section and the individual pest profiles feature key biology, physical and behavioral characteristics.

PCT Field Guide for the Management Of Urban Spiders

By Stoy Hedges and Dr. Mark Lacey

An essential addition to every PCO's technical library, the spider field guide covers basic spider anatomy, behavior and biology points, keys to proper identification and tips for implementing a proper spider management program. From active hunters to web builders, the book offers detailed profiles of the most commonly encountered spiders and features more than 100 illustrations. The health aspects of spiders is also covered in this information-packed publication.

PCT Field Guide for the Management Of Structure-Infesting Beetles

By Stoy Hedges and Dr. Mark Lacey

This handy, pocket size guide to the biology, behavior and control of beetles is written in easy-to-understand language featuring information on key biological, behavioral and physical characteristics of hundreds of structure-infesting beetles. The comprehensive field guide is divided into two volumes with Volume I featuring Hide and Carpet and Wood-Boring Beetles, and Volume II featuring Stored Product and Overwintering Beetles. Both volumes contain hundreds of detailed illustrations and color identification sections.

Handbook of Pest Control 8th Edition

By Arnold Mallis

Edited By Stoy Hedges

For a half-century, pest control professionals have turned to the Mallis Handbook of Pest Control when they've had questions about the biology, behavior and life history of structural pests. No other publication offers the depth of information contained in this 1,500-page desk reference book that is essential for any industry professional. The recently revised and totally updated 8th edition includes editorial contributions from 30 leading industry experts and includes more than 1,000 photographs and insect illustrations. Also included are comprehensive insect keys, color photograph sections, a glossary of technical terms and a pesticide reference chart.

For Pricing And Ordering Information Call The PCT Book Department At 800/456-0707

About The Author

A well-respected name in pest management education, Dr. Richard Kramer is president of Innovative Pest Management, a pest management consulting firm based in Olney, Md. A Board Certified Entomologist, Kramer also serves as director of technical services for American Pest Management, Takoma Park, Md. A former technical director for the National Pest Management Association, he is also contributing technical editor for *Pest Control Technology* magazine. He resides in Maryland.